The BRC Global Standard for Food Safety

T0176276

The BRC Global Standard for Food Safety
A Guide to a Successful Audit

Second Edition

Ron Kill

Micron² Ltd
Market Drayton
Shropshire
UK

A John Wiley & Sons, Ltd., Publication

This edition first published 2012 © 2012 by John Wiley & Sons, Ltd.
First edition © 2008 by Ron Kill

Wiley-Blackwell is an imprint of John Wiley & Sons, formed by the merger of Wiley's global Scientific, Technical and Medical business with Blackwell Publishing.

Registered office: John Wiley & Sons, Ltd, The Atrium, Southern Gate, Chichester, West Sussex, PO19 8SQ, UK

Editorial offices: 9600 Garsington Road, Oxford, OX4 2DQ, UK
          The Atrium, Southern Gate, Chichester, West Sussex, PO19 8SQ, UK
          2121 State Avenue, Ames, Iowa 50014-8300, USA

For details of our global editorial offices, for customer services and for information about how to apply for permission to reuse the copyright material in this book please see our website at www.wiley.com/wiley-blackwell.

Library of Congress Cataloging-in-Publication Data

Kill, R. C.
   The BRC global standard for food safety : a guide to a successful audit / Ron Kill. – 2nd ed.
      p. cm.
   Includes bibliographical references and index.
   ISBN 978-0-470-67065-1 (pbk. : alk. paper)   1. Food industry and trade–Standards–Great
Britain.   2. Food industry and trade–Standards.   I. Title.
   TX537.K54 2012
   338.1'9–dc23

                                                              2012010600

A catalogue record for this book is available from the British Library.

Wiley also publishes its books in a variety of electronic formats. Some content that appears in print may not be available in electronic books.

Cover design by www.hisandhersdesign.co.uk

Set in 10/12pt Sabon by Aptara® Inc., New Delhi, India
Printed and bound in Malaysia by Vivar Printing Sdn Bhd

1 2012

Contents

About the Online Training Resources

BRC Update Course Issue 5 to Issue 6

This course guides the delegate through all the major changes of the BRC Global Standard for Food Safety from Issue 5 to Issue 6 in an easy-to-use e-training course. The course has been developed to complement the book and provides a searchable database of all the changes.

To start using the course, just email info@micron2.com and request your username and password today, stating 'BRC Global Standard Course' in the subject line.

For more details on how to start using the course in conjunction with this book, just go to www.wiley.com/go/kill/brcglobalstandard.

Micron^2 in conjunction with **totrain** are also providing other food-related courses; for more information go to the Micron^2 web site at www.micron2.com and click on the e-learning link.

Cast of Characters

Accreditation Body: the body that bestows accreditation on to Certification Bodies to enable them to operate the scheme. They feature from time to time.

Approved Training Provider: an individual who is registered with the BRC to carry out their official training courses.

Auditor: with a capital 'A' this refers to the person carrying out the audit for the Global Standard for Food Safety (as opposed to your internal auditors). They are central characters in this book.

Certification Body: the Company that bestows certification on manufacturers and who employs Auditors to carry out the Audits. They also feature frequently.

Clause: this term is used to describe the individual requirements of the Standard.

Codex Alimentarius: there is only one document referred to specifically in the Standard. It is Codex Alimentarius HACCP Principles (Basic Texts: Hazard Analysis and Critical Control Point (HACCP) System and Guidelines for its Application). It is used as the basis for the Food Safety and HACCP requirements here. I sometimes shorten the name to simply 'Codex'.

Paragraph: this term is used to describe the numbered sections of the Protocol.

Stakeholders: these are the people, mostly members of the BRC, who instigated the process of third-party certification and who have a large say in the continuing process and the Standard reviews. They usually require sight of copies of Audit reports of their suppliers. They appear mainly as background characters here.

You: In this book, I refer to 'You' and 'Your' a lot. Note that the book is intended for manufacturers seeking certification. I know some Auditors find it useful too but here 'You' refers to all readers who seek certification.

Abbreviations

Being a dis-liker of jargon, I resist the use of abbreviations, especially acronyms. However, for the sake of brevity, I have used those as follows.

AB   Accreditation Body
ATP   Approved Training Provider (see above) or adenosine triphosphate
           depending on the context
BRC   British Retail Consortium: a member-based body in the UK largely funded
           by UK retailers.
BSDA   British Soft Drinks Association
CB   Certification Body
CIP   Clean in Place
EFK   Electric Fly Killer
FIFO   First in – First out
FSA   Food Standards Agency: a public body set up in the UK to give
           independent information on legislation, nutrition advice, food safety
           messages and alerts.
FSQMS   Food Safety and Quality Management System
GFSI   Global Food Safety Initiative
GM   Genetically Modified
GMP   Good Manufacturing Practice
NPD   New Product Development
SOI   Statement of Intent
UKAS   United Kingdom Accreditation Service: the AB that functions in the UK
WIP   Work in Progress

Acknowledgements

My heartfelt thanks to Adrian Davies, who was a great help in researching examples and correcting my grammar for the first edition of this book. Thanks to David Brackston of the BRC for allowing me to use his nonconformity decision tree and for answering all my questions from my Issue 6 courses. Thanks also to the BRC itself for the use of their web pages in this book.

Thanks to my wife Eve for putting up with all this – again!

And as always to the memory of Michael.

Introduction to Second Edition

It seems a long time since I drafted the first edition of this book, which was originally intended to be a companion to Issue 4 of the Standard. In the end, it appeared alongside Issue 5. In terms of this industry, of course, 5 years is a long time and much has changed, not least the structure of the Standard. Issue 6 has some very new ideas, borne partly out of some criticisms of the nature of the Audits, partly out of the need to move on in terms of the value that the Audit offers the company.

The original introduction to the first edition is included again here but please note that the references to Issue 5 matters are largely irrelevant. I kept it in for historical interest and some of the ideas still apply. This book is all about Issue 6.

For this edition, I have retained many of the real-life examples of nonconformities from the first edition as they are still relevant. However, I have changed the clause numbers to match the current issue of the Standard. I have also added a good number (about 200) of new examples. I have changed the structure of the book so that it moves forward in the same way that the Audit process moves. In doing so, I have added a few new chapters, including ones on the BRC Directory and aspects of Training. The structure of this edition partly reflects my time spent on the working group that reviewed the Standard and the training for Auditors that I helped with. I will freely admit that I gained much information and ideas from delegates on courses.

My aims remain the same though – to help the manufacturer through what might feel like a step into the unknown. I also hoped to make it as readable and entertaining as possible. It can be a dry old subject after all.

Sir Steve Redgrave said after winning his fourth Olympic gold medal: 'I've had it. If anyone sees me near a boat they can shoot me.' (He subsequently went on to win gold – again – at the Sydney Olympics). I am not exhausted like he was when he said it but at the risk of sounding like Sir Steve and then eating my words, I have said that I will not take part in the next review in 3 years time. Assuming I am still around I will be 62 and I think it is time for some fresher minds to take over. So, I doubt I will be going for gold again, but you never know.

Ron Kill
January 2012

Introduction to First Edition

It is Friday afternoon, the phone rings. It is your National Account Manager:

> Look, we have just got the contract with (insert name of major retailer). It's brilliant, should double our turnover but now it's over to you 'cause we have to get the BRC Standard and we need to be up and running in 12 weeks. OK?

Sounds familiar? Apart from ruining your weekend, what goes through your mind? A sinking feeling that you are alone and that you have a mountain to climb? Or a feeling that this is no problem because you have the comfort of a large team around you and the knowledge that your systems are sound. Either way, this book is intended to guide you through the process so that come the days of your Audit you are prepared in the best possible way.

The Global Standard for Food Safety was originally conceived in the mid-1990s when there was an increasing demand for a unified standard to be used by the major retailers in the UK for their suppliers of 'own label' food products. I think it is fair to say that up to that time UK retailers were, certainly in Europe, among the leaders in demanding good standards of their food suppliers.

As always with new initiatives, the gestation was quite long, but the first BRC Standard was published in October 1998 with the title: The British Retail Consortium Technical Standard for Companies Supplying Retailer Branded Food Products. *(I am reminded of an old Peanuts cartoon in which Snoopy is trying to write his great novel but cannot get past the title because all the good ones: 'Gone with the Wind', 'For Whom the Bell Tolls', 'Of Human Bondage' are used up).* In this case, the title did have some real meaning because the idea was to protect those retailers selling 'own label' products by ensuring the good standards of their suppliers. What rapidly happened, however, was that it was successful in all areas of the food industry not just in private label retail supply, so that during the life of Issue 3 it was renamed to reflect its 'Global' status. In this book, I shall refer to it as the Global Standard or simply the Standard.

Issue 5 of the Global Standard for Food Safety was published in January 2008 and introduced some significant changes to the Certification process, all of which are dealt with here.

The protocol and requirements discussed in this book are based on Issue 5.

Supermarkets in the United Kingdom and around the world have been part of an economic shift over the years in driving down the prices of food. We all now spend a fraction of the proportion of our disposable income on food that our grandparents did but at the same time food is manufactured in volumes and is of a consistency that would not have been possible in those earlier times. Food suppliers have had to become efficient places who can only survive if every penny is watched. Technical staff are sometimes seen as a luxury. My aim in writing this book is to make life easier for the QA and Technical Managers of suppliers who have to prepare for their annual Audits while struggling with all the day to day problems of running the technical and quality side of their business.

In this book, I stress the necessity of having good systems in place (which meet the Standard) all the time. This is important. As a supplier you should be endeavouring to have systems and records in place which meet the Global Standard all year round. Your certificate should not be seen as the objective but the outcome of having good standards that benefit you as well as your customers.

However, it is also important to be able to demonstrate this to the Auditor and for you to be well prepared for the Audit. In our example, a 12-week lead time is not to be recommended, to be honest. As I explain in the book, systems not only have to be in place but you will have to demonstrate that they are working and produce evidence of a good track record.

We would not be human if we did not want to present ourselves in the best possible way when it counts most. So, having good systems in place may count for little if you are poorly prepared for the Audit and are unable to convince the Auditor that you have got it right.

I also aim to make life better for all those intrepid Auditors in the field who may be struggling to explain what it is they need to see or wasting valuable time while they wait for the supplier to present the right document during an Audit. Being an Auditor is sometimes perceived as something of a glamorous life. Yes, there are great moments and for anyone who loves being in factories as I do, it is interesting and instructive. You can meet wonderful people and it can be very satisfying, especially seeing a site improve over time. There is a down side though. It is exacting work, the days are long, travelling is tiring. However, in general, it is manufacturer's lack of preparation for the Audit that makes the Auditor's life most difficult. If suppliers are better prepared for the Audit, then the Auditor's life will be better too.

I have worked for major retailers and have been in both manufacturing and importing businesses to supply them. I now sit in the middle, auditing suppliers in a system that is on behalf of the retailers and other purchasing groups. I know that for the most part both sides of the supply chain are trying to do a good job with the best of intentions. I want to help suppliers to understand the reasoning behind the clauses of the Global Standard for Food Safety, to perform better at their Audits and in doing so give their customers the assurances they need.

Ultimately, I have tried to let you see what might be going on inside the Auditor's head so that you can be better prepared. Where possible, I have illustrated points with examples, usually of where the Auditor has found something wrong. This may seem slightly negative but my intention is that you take something positive from it.

Every 'Example' that I use in the book is real. Of course, no names are mentioned. If you think you recognise yourself either (a) you are probably wrong because most of these examples repeat themselves quite regularly at other suppliers or (b) you are right but consider (a) and remember it is in a good cause.

Auditors are human beings (trust me). In this book, the intention is to give you a good idea of what an Auditor is likely to want to see. Of course, everyone has their own ideas, experiences and personality. I cannot tell you exactly what the Auditor will ask for on the day but I hope that I have covered enough ground and given sufficient examples to prepare you for all likely situations.

A note on definitions. For issues 3 and 4 of the Standard, we became used to using the terms 'Evaluation' and 'Evaluator' to describe the process of auditing against the Standard and the person carrying out the audit. With Issue 5, it has been decided to revert to the more universal terms 'Audit' and 'Auditor'. I now have a slight regret about this because I did find it useful to be able to distinguish between this activity and your own internal auditing by using different wording. For the purposes of this book, therefore, I will use the capital 'A' to mean the Certification Body's (CB) activity and lower case to indicate your own.

The factory or production unit receiving an Audit is described throughout mostly as a 'company' although I also use the words 'manufacturer' and 'processor' and similar expressions in my examples. The word 'supplier' is reserved for suppliers of raw materials, packaging and services to the company being evaluated.

The Standard has a number of requirements set out into 7 Clauses. In fact, in Issue 5, there are 325 clauses in all including the Statements of Intent. Although the proper term here is 'requirement' I often use the word 'clause'. Again, for the purposes of this book they mean the same thing. Similarly, I also use the term 'sub-clause' quite frequently when referring to a requirement that is not an emboldened heading in the Standard and usually bears a string of numbers such as 4.3.2.2.3. They are all individual requirements.

Regarding the structure of the book, I have divided it into three parts. The first part is about the company's general background knowledge and to prepare you in general terms, including the important aspects of the Protocol. In the second part, I look at every clause of the Standard (all 325) and describe what the Auditor is likely to be looking for.

In the third part, I look at what should happen in the days after the Audit leading up to, hopefully, certification and beyond.

Although I discuss every clause in the Standard, for the sake of brevity, I have not reproduced the exact wording of every clause. This book is intended to be read with a copy of the Standard at your side.

Good Luck.

Ron Kill
March 2008

Part One
Before the Audit

In this part of the book, I give you all the background information you need on the decisions you must make and the procedures you must follow prior to your first audit.

The chapters are:

1   The Changes: Issue 5 to Issue 6

If you are new to the Global Standard, then this chapter may be only of passing interest and you may want to skip to Chapter 2. However, for those who have been involved before or if you are just interested in the history, here is a look at the key headline changes in the Standard from Issue 5 to Issue 6.

Reasons for Change

There were four big issues with the stakeholders for the review this time around:

(1) Firstly, it had been felt for some time that too great a proportion of the audit time was being spent by the Auditor looking at documents and becoming trapped in the office when they should be where the action is: in the factory. Indeed, some were saying that audits were not picking up issues of housekeeping or building and plant fabric because not enough time was spent in production. Furthermore, it was felt that potential weak points in production such as shift changes and product changes should be audited, and that the Auditor should aim to be present at those times.
(2) Secondly, there was disappointment in the low uptake of the unannounced audit scheme, which for Issue 5 attracted less than 100. They were looking for a change to encourage more to try unannounced but thankfully drew back from making it compulsory.
(3) Thirdly, there was a desire to see more new companies going for the Standard, especially in developing countries, which might be put off by some of the Requirements.
(4) Fourthly, there was a desire to nail the terms 'high care' and 'high risk' and bring better understanding of these terms.

Other changes were made this time as a result of the wide consultation with CBs and interested parties in industry.

The BRC Global Standard for Food Safety: A Guide to a Successful Audit, Second Edition. Ron Kill
© 2012 John Wiley & Sons, Ltd. Published 2012 by John Wiley & Sons, Ltd.

Two-Part Auditing: Increased Focus on GMP

The challenge was how to allow the Auditor to spend more time in the factory. The answer was to have an overall aim of audit time being split 50:50 between production areas and document review in the office and to achieve this by both the review itself and the resulting format of the Standard and by training. *In my opinion, this is the most significant change for Issue 6 and will result in a different feel to the audit for all involved.*

To help with this, we have now a Standard where the Requirements are divided by colour coding to indicate to the Auditor that which should be audited in each place. In general, GMP aspects will be audited during the site tour (see Chapter 10).

There is also a significant aspect of training in this idea, and the official training courses for Auditors will now include emphasis on spending more time interviewing staff, collecting evidence in the factory, observing product change times and so on (see Chapter 10).

Vertical Audits

To add to this change, we now have more emphasis on what we term as a vertical audit. This will be done in parallel to the traceability challenge carried out by the Auditor. It will be a document review but will take in all the process records related to the traceability exercise being done. Furthermore, it will include all related issues starting with raw material specifications and supplier approval right through to dispatch and finished product release, taking in any relevant training records etc. on the way. In this way, this will be a complete vertical slice through what you do and a valuable way for the Auditor to sample much of your documentation.

Unannounced Audits

We now have two options. The first is mostly as before with some changes, mostly relating to timing and dating. The new Option 2 means that you can opt for an audit that is part unannounced, part announced. The unannounced part will be a GMP audit of your site, mostly spent in the factory. The second part will be *announced* and take place before the due date as usual. It is hoped that this will encourage more to choose an unannounced option. I am interested in your views, so email me (ronkill@micron2.com).

Enrolment Programme

Essentially, this idea is to encourage new starters to join the scheme with no fear of 'failure'. There will be no stigma of failing to achieve a grade but rather a score given that can indicate progress towards certification. As a consequence, this is now the entry point for all companies into the Standard and we will no longer apply the term 'initial audit'. However, a company may stay in the Enrolment Programme for as long as necessary. One incidental result is that there is now no 90-day grace for resolving a major nonconformity at an initial audit. The thinking being that if you cannot correct a nonconformity in 28 days, you will be un-certificated and remain in the Enrolment Programme or rather the continual development phase (see Chapter 5).

The Protocol itself is also now in two parts, the second of which details the options for the format of audit now available to companies.

High Care and High Risk

New definitions of these terms have been written into the Standard but on the basis of areas or zones in the factory, not the product. The Standard now has an Appendix (2) of five pages covering all levels of risk as applied to the factory areas. There are also some new and rewritten Requirements as a consequence (see Chapter 7).

New Report Format

A new style of report has been developed, with less detail on each clause but a better summarising of each section. There is also more useful detail in the front pages of the report on the company (see Chapter 18). It is hoped that this will be more readable and it will certainly be better for the Auditor to write.

Other Protocol Changes

No more Grade D

The term 'No Grade' will be used instead in the appropriate grade box on the report and 'Not certificated' in the audit result field as before.

No factored goods

Only products made at the site can be included in your scope.

Root cause

For every nonconformity, you will not only have to submit evidence of your immediate corrective action, but analysis of the root cause of any issue and plans to address it. This includes nonconformities from your standard audit and from any internal issues identified. For more details on root cause analysis, see Chapters 4 and 19.

Head office audits and multiple locations

There is now a system for separate audits of head office functions, where appropriate. There are also new requirements for the auditing of multiple sites and off-site storage areas (see Chapter 16).

Changes to Requirements

There are numerous changes to the Requirements.

Many clauses have been merged, so there appears to be fewer than last time. The total is down from 325 to 284, including SOIs. Do not be fooled; there is plenty of material there. One of the reasons for merging some of the old clauses was to even out the weighting of clauses. A consequence is to make it more appropriate for a minor nonconformity to be given when a certain issue is not complied with.

Some clauses have fallen by the wayside and in one case an entire subject: *customer focus* has been removed, for example. No one was sorry to see that one go.

Many clauses have been extended to be more prescriptive and easier to understand. Thus, more clauses have lists of specific points to meet.

Other changes include greater emphasis on prerequisite programmes in HACCP and more reference to agency workers in personnel.

Statements of Intent

SOIs have been revised and almost all have been rewritten so that they truly are objectives and not specific requirements in their own right.

Note: Appendix 2 of this book gives details on where all the Requirements of Issue 5 have gone and Appendix 3 gives new clauses for Issue 6.

2 Keys to Success

In this chapter, we look at success and failure and some basic ideas to get you tuned into the right way of thinking about all this so you don't have to worry about that f-word (failure).

Key points in this chapter
Success and failure
Preparation
Documents and evidence
Track record
Team building
Internal audit

Success

How do we measure success in terms of the Global Standard for Food Safety? Naturally, your first thought will be to get your certificate, as your passport to meeting your customers' requirements and being able to continue to supply them, and hopefully increase your business by finding new customers.

Your second thought though should be this: the Standard is a system that requires and encourages your continual application and attention throughout the year. As such, the real success is not about that nice piece of paper in a frame in your reception area; it is about meeting the Standard for the 363 days in a year when you are not being audited, because the real result of this achievement is that you will have an excellent management tool for ensuring the safety and legality of your products.

You should aim to be confident in your systems so that the audit is not a trial but a challenge that you expect to meet well. Truly, that would be success. You might even consider an unannounced audit as the ultimate challenge, but even this you should be able to approach with confidence. For information on unannounced audits, see Chapter 5.

The BRC Global Standard for Food Safety: A Guide to a Successful Audit, Second Edition. Ron Kill
© 2012 John Wiley & Sons, Ltd. Published 2012 by John Wiley & Sons, Ltd.

The grading system for the Global Standard is discussed in Chapter 18. Essentially, a certificate will bear Grades A, B or C (or A+, B+, C+ if you have an unannounced audit). Remember that the Global Standard is a demanding and exacting one. It is highly likely that even the best prepared site will receive some nonconformities because of a fresh pair of eyes looking at it, and you should not be disheartened. Furthermore, success does not mean only Grade A (or A+). All the grades from A to C are an achievement because it means that any issues found were not so great or numerous that you could not be certificated and you have been able to correct all the nonconformities given.

Success is also about continual improvement: looking at yourselves and, through your review systems, finding ways to improve. This is so important. A static system will inevitably start to stagnate and this will ultimately lead to you falling behind your competitors in terms of efficiency and customer satisfaction.

Failure

Like the old joke about nostalgia, failure is not what it used to be. Whereas in the past we would consider that failure is not achieving a certificate, now we look in a more positive light at any company joining the scheme, and even if you fail to achieve a certificate, you will now be given a score under the Enrolment Programme (see Chapter 5). This is a step away from the hard and fast rules that we had in the past for initial audits; effectively, for an initial audit there is no failure, only a score.

If you are already in the scheme, you might fail to achieve certification, and in this case you will not get a certificate or a score. The most likely way this can happen is if you are issued with nonconformities during the audit to such a degree that you do not meet the criteria for an A, B or C grade. For example, if you fail to meet a Fundamental Requirement with a critical or major nonconformity or if you have a critical nonconformity against any clause, you will be uncertificated. Other combinations of major nonconformities against other clauses would also result in being uncertificated.

Another way to end up being uncertificated is if you fail to correct any nonconformities in the required timescale (28 days). This could be the result of poor organisation or lack of resources, or again, poor preparation, or quite simply that you only aspire at this stage to be in the Enrolment Programme.

So, what are the keys to success?

Embrace the Standard

Well, all right, I don't expect you to love it. But again, the Standard works better for you if you take it on as a philosophy and a core principle to running the business. Try to imagine that you will be audited that very day. What effect does this have on your daily approach to the business? Also, try to instil this feeling in your colleagues, including senior management and operatives. The best companies to audit and the ones that perform better have an air of having placed the Standard at the heart of their thinking. There is often a positivity that shines through.

Preparation

Preparation is the most important word in this book. It is the reason you are reading it, so a good start!

The first practical thing that you must do is to obtain a copy of the Standard. You can buy it online at www.brcglobalstandards.com or through TSO at www.tso.co.uk. It is available in various languages and in some cases in electronic format (as a .pdf file) as well as hard copy. This book should be read with the Standard at your side. In Chapters 5 and 18, we look at all of the very important matters in the Protocol for the Standard. *I cannot emphasise enough how important it is to be familiar with that part of the Standard and that you should not just focus on the Requirements section.* In Part 3, we look at each requirement clause of the Standard in detail.

This book will help you to go through all the various stages to prepare you, but in essence, you will need to have systems and procedures in place and ensure that you can readily demonstrate them to the Auditor and that they are working. This will mean not only adhering to your systems but also being able to show that you do so with documents and other hard evidence. It is easy to read the Requirements and imagine that you can meet every one. But it is very rare for even a first-class operation not to have some nonconformities. Often, the reason for that is lack of preparation for the audit.

Example

A yoghurt factory was hoping for a Grade A certificate, but they have been awarded a major nonconformity because they could not show the Auditor a documented supplier approval procedure. Later that night, after the closing meeting, the Technical Manager found a procedure on his computer at home and called the CB frantically next day to say he can now provide it; it was honestly there all the time and can the CB remove the nonconformity?

Unfortunately, they cannot as it was not in place during the audit and certainly not available for other staff to see. His major nonconformity remained and the site was later awarded a Grade B certificate.

Documented Procedures and Records

While on this subject, it is worth pointing out that there are many clauses in the Standard that ask for a procedure or a work instruction. They must all be documented. The word 'documented' appears in the Standard 84 times. It is now the case that *all* procedures must be documented in some way, as described in the Glossary to the Standard. It would also be expected by the Auditor that records are also documented. Therefore, as part of your preparation, you must ensure that all necessary systems are documented and that your documents are controlled. We discuss controlled documents later in Chapter 13.

Evidence

Here is another key word, especially for the Auditor. The Auditor is looking for evidence to support the fact that you meet a requirement. The Auditor has to record all the evidence that they find in order to support the fact that you meet the requirement. Your preparation should include making sure that you have that evidence, whether it is a physical, tangible thing that the Auditor can see or whether it is a document or record. You must also have the evidence readily available.

It may be that the Auditor wants to test the retrieval of your records. For example, you might say in your procedures that you maintain records for 3 years; so be prepared for the Auditor to ask for a 3-year-old record. Failure to be able to do this and meet your procedure might result in a nonconformity.

As far as documentary evidence is concerned, increasingly much of this is now seen in the form of computer records as well as paper records. There might be nothing that pleases some Auditors more than good paper records with 'real' signatures, but it is understandable that many manufacturers want to keep electronic records. In this case, the Auditor will be looking for some assurances of authorisation of these records and probably systems for limiting access to certain personnel, password protection, backing up, audit trail and so on.

Part 2 will deal with examples of the kind of evidence you will need for each section of the Standard.

Track Record

In the original introduction to this book, I indicated that 12 weeks was on the tight side in terms of preparing for the audit from scratch. Indeed, 3 months is considered the very minimum time for preparation for an audit. If you already have reliable systems in place and have been keeping good records, then, yes, it would be possible; however, the key phrase here is track record. The Auditor must have confidence that what they see is representative of your continuous controls and not a 1-day wonder to get you through the audit.

For a completely new site, I would say it is not possible to audit without at least 3 months' track record, and preferably more. Indeed, the Protocol paragraph 7.1 indicates that it is unlikely that full compliance can be demonstrated with less than 3 months of operation.

Team Building

When you are establishing a HACCP system, you are encouraged to have a team involved, preferably from several disciplines within the factory. If you are a Quality Manager or Technical Manager, a team approach to the whole Global Standard is a good one, not only to share the burden and spread the load but also to ensure that systems are better maintained. You will be expected to have other members of management involved in the Quality System and take part in system reviews, internal auditing and so on; so it is a good idea to make your ongoing systems a team effort.

In a small company, your team may only consist of two people, but even so this will pay dividends. For the Auditor who arrives to find that only one person on site knows anything about the Standard, alarm bells will ring from the start.

A consultant may form a useful part of your team, especially where resources are limited and your staff number is small. By all means, consider this option if you feel unsure of the prospects for your audit. A good consultant can make all the difference to the outcome for you, especially if you are lacking in resource in your preparation. The expertise and experience they can bring to your team can be considerable. However, choose them with care and make sure that they are team players. They may also play a significant part in the audit itself and represent you well. On the other hand, their very independence can sometimes cause problems as I discuss in Chapter 10.

Internal Audit

The importance of internal audit of your systems cannot be exaggerated.

It is essential that you are carrying out internal audits for some time before your audit. There are two reasons for saying this.

Firstly and most importantly, it is the only way you can be sure that all your policies and procedures are being carried out in the proper way and in the ways in which you have set them out. Done properly, internal audit should highlight any deficiencies in procedures and practices and give you the opportunity to put things back on course long before the BRC Auditor gets a chance to do the same.

Secondly, you need to do it to satisfy a Fundamental Requirement of the Standard (Clause 3.4), and in order to convince the BRC Auditor, you will need to be able to present enough history, records and so on to demonstrate that it is a working system. For reasons that are expressed later, you will need to consider the competence and training of staff involved in internal audits. You will need at least two auditors to ensure independence of audit.

See Yourselves as Others See You

O wad some Power the giftie gie us
to see oursels as others see us!
(Robbie Burns, to a louse, on seeing one on a lady's bonnet at church)

Try to imagine what it is like for a complete outsider to understand your thoughts and philosophies, your policies and practices in a couple of days. This is especially the case when an Auditor is with you for the first time. You may have been in your situation for many years and have grown and evolved to a point where you have a well-run operation. But guard against complacency and take nothing for granted. Spend some time considering how you can best present yourselves to someone else, in what is only a snapshot of what you do.

To use the example of documented procedures again, the very worst thing to say to an Auditor who asks to see a documented procedure is to admit 'well, actually we don't have it written down because that is always done by [person's name], and they have been doing the same job for absolutely years'. There are many arguments against this approach such as 'how would you train new staff in procedures if they are not documented?' or (my personal favourite) 'so if the Production Manager drops dead tomorrow, does anyone else actually know what to do?'

Sometimes, you have to take a mental step outside your day-to-day environment to see how you could justify your approach to an outsider, and don't miss spotting the louse yourself!

Try This

Take an aspect of your systems, an ambient storage area for example, and go and look at it *as if for the first time*.

Just stand there and take it all in for a while. Perhaps take some photographs.

Then look closer at the following:

- Floor–wall junctions. (Are they clean? Any stray rodent bait? Any damage?)
- External doors. (Are they closed? Are there gaps at the base?)
- Products. (All off the floor? All away from the walls? Allergens segregated?)
- Pallets. (Condition?)
- Racking. (Condition?)
- Lighting.
- Allergen segregation.

I could go on. You can make your own list, but try to see everything afresh. Then try this out with a different aspect, a system-like control of a CCP, for example. Again, go back to basics in your mind and give it a fresh overview, then a detailed examination.

Another way: Get a colleague from a different area of the business to do it for you.

Summary

In this chapter, we have had a brief run through the kind of mindset you should have to ensure success at your audit. Hopefully, this will set you in the right frame of mind for the rest.

Quiz No. 1

Which Fundamental clauses are mentioned in this chapter?

3 Some Background

Later in Part 1 we look at various aspects to consider in the run-up to your audit, but before that, here is some background information about how CBs work and how we all got here and the systems in place to ensure consistency. You don't need to know all this stuff, but you should find it interesting.

Note: A plea on behalf of all CBs: please read the paragraph on witnessed audits!

Key points in this chapter
CBs and ABs
Consistency
Co-operation Groups
Witnessed audits
The BRC's KPIs

What is a Certification Body?

A CB is a business that has been accredited to audit against and issue certificates for compliance with specific standards. To get to that point, the Standard itself has to be recognised by the AB concerned and deemed accreditable. Because the Global Standard was created in the United Kingdom, it was originally recognised by the UKAS, then by all other ABs around the world.

Note: What CBs do is certification, not *accreditation*. The BRC system is not accreditation but certification.

Accreditation and Competency

The accreditation process that the CBs have to go through is in some ways similar to the certification process that you are planning to go through. They have to be accredited against an international standard called EN45011, which means they are accredited to act as CBs for Product Certification. In each country, there is one AB that may bestow this on the CBs; for example: in the United Kingdom, it is UKAS; in Italy, it is Accredia; in Norway, NA; in South Africa, SANAS and so on.

The fact that CBs are accredited for Product Certification has some significance for you. In the early days of the BRC Standard, before Issue 3, a different international standard was imposed on CBs (EN45004). In fact, they were Inspection Bodies rather than CBs. Now as CBs, it is clear that your products are being certified, which means that the scope of the audit that you agree upon must be very clearly defined. You must also specify those products not included in the scope (see Section 'Agreeing the Scope', Chapter 7).

In addition to meeting EN45011, CBs also have to comply with a set of requirements issued by the BRC called 'Requirements for Organisations Offering Certification Against the Criteria of the British Retail Consortium Global Standards'. Once again, it has an unwieldy title; so from here, I will refer to it as 'Requirements for CBs'. If you can get to see a copy, I recommend that you do read this, as there is very useful information for you in there. (It is not available in the public area of the BRC website.) Some of this document, notably the qualifications and training requirements for the auditor, now appear in the Standard as Appendix 3.

The 'Requirements for CBs' document imposes a structure such that the food industry is divided into six fields and sub-categories. The six fields are listed as follows:

(1) Raw products of animal origin that require cooking prior to consumption.
(2) Fruit, vegetables and nuts.
(3) Processed foods and liquids with pasteurisation or UHT as heat treatment or similar technology.
(4) Processed foods ready to eat or heat foods, that is, heat treatment or segregation, and processes that control product safety.
(5) Ambient stable products with pasteurisation or sterilisation as heat treatment.
(6) Ambient stable products not involving sterilisation as heat treatment.

Schedules of accreditation

In going for accreditation, the CB has to declare to the AB which of these fields they wish to be accredited for. When accredited, the CB is issued with a Schedule that includes all Standards it is accredited for and, in the case of BRC Standards confirming those fields, it is accredited to audit as a CB. If the CB is a UKAS-accredited one, you can view these schedules on the UKAS website at www.ukas.com. Otherwise, the CB should be able to send you a copy of their schedule on request.

Thus, you can confirm that the CB you wish to use is indeed accredited in your field of production. In the case of CBs still seeking accreditation, you could ask to see the relevant parts of their Quality Manual and a copy of a letter from the AB confirming the fields they are going for. For more detail, see Section IV of the Standard: Management and Governance of the Scheme.

Auditor competence

The fields numbered 1, 2, 4 and 6 are further divided into sub-categories to produce a total of 18 product categories. For example, 6 is divided into beverages, alcoholic drinks and fermented brewed products, bakery, dried foods and ingredients, confectionery, breakfast cereals and snacks, oils and fats. The Auditors employed by the CB to carry out audits have to demonstrate competence in a category before they may audit such a company.

Note: There were some minor changes to the product categories for Issue 6. These are tabulated in Appendix 4 of this book.

So not only does the CB have to demonstrate overall competence in the six fields it is accredited for, the Auditors have to demonstrate competence in the particular product categories they will be auditing. This might influence your choice of the CB in that you need to be sure that they have the resources to match your product range and are not only accredited in your field but can offer an Auditor in your specific category and in your desired timescale.

See also Appendix 3 of the Standard for the qualification and training requirements for Auditors.

A CB's Contract with the BRC

The requirements on CBs are very exacting. In addition to all the aforementioned requirements, the CBs must be registered and have signed a contract with the BRC. The contract imposes further requirements on them, including (you will be comforted to know) a sizable level of professional indemnity insurance. They are also required to collect and administer the BRC registration fee and provide information on the status of companies to the BRC. Furthermore, the BRC themselves now monitor the performance of CBs very closely and will suspend a CB if they do not meet their requirements (see Section 'BRC Competence Monitoring').

Consistency

Co-operation Groups and Position Statements

The nature of CBs varies from quite large companies with many staff to small businesses with a handful of Auditors, some of which may be contractors. Your choice may depend in part upon which type of company you are more comfortable dealing with. However, one of your concerns may be consistency of approach between them regarding the interpretation of the Standard.

Here, I can give you some assurance. Some time ago, about the time that Issue 3 of the Standard was introduced, there was concern among the UK CBs about consistency. As a result, a group was formed by the UK CBs called the FTCG. This group meets regularly about five times per year and one of the aims is to discuss clauses of the Standard and how we interpret them. The meetings are attended by the CBs operating in the United Kingdom, and any other accredited by UKAS and by representatives of the BRC and UKAS. They are very useful and productive meetings in which the CBs can air their issues or difficulties in applying certain clauses, feedback from client companies and so on. Sometimes, the discussions on interpreting a clause will result in a formal Position Statement, which we all adhere to. After further discussion with the BRC at the TAC, these are published on the BRC website. Some of these Position Statements are available in the public area of the BRC website, some are only available in the members area (see Chapter 8).

The great thing about the Co-operation Group idea is that it is driven by the CBs, and not by any of the BRC or AB hierarchy, and demonstrates just how much we do care about our own standards. As one of the founding members, I am also quite proud to see that the BRC were so impressed with the idea that they now recommend it to all countries and have written the concept into the Standard. It may come as a surprise that business competitors can sit down and co-operate in this way, but it works well. Indeed, there is now an annual International Co-operation Group meeting held at the BRC Conference.

Furthermore, when other BRC Global Standards were launched, such as for packaging, the Co-operation Group principle was firmly encouraged by the BRC and ABs right from the start. Representatives of the Co-operation Groups in turn attend the regular TAC meetings of the BRC so that there is always a good two-way dialogue between the parties involved.

A look at these Position Statements will hopefully convince you that the CBs not only care about consistency to a significant degree, but that there is real concern among them about the proper application of the Standard.

Accreditation bodies

The ABs also play their part in consistency during the accreditation of CBs and subsequent surveillance processes. Each CB is assessed annually by the AB. This involves an office-based assessment of their systems and procedures and at least one witnessed audit on site. Naturally, their judgement of CBs must itself be consistent. What about consistency across the world? This too was of concern a few years ago when there was a rapid growth of interest in the BRC Standard around the world. Quite suddenly, CBs in many countries outside the United Kingdom were looking to their national AB to accredit them to operate this Standard and there was initial concern about the interpretation by those ABs of both the BRC Standard and how EN45011 applied to it.

A great deal of work was done by UKAS and the BRC to tackle this potential issue. The BRC has also issued a document entitled 'Guidance For Accreditation Bodies Assessing Certification Bodies for BRC Technical Standards', which is a companion document to the 'Requirements for CBs' document referred to in Section 'Accreditation and Competency'. Again, this document is aimed at achieving a consistent approach at this level so that the CBs are all judged in the same way. Agreements have been reached with ABs around the world on the interpretation of EN45011 with regard to accrediting the CBs and in this way the best possible job was done to ensure that there was consistency of approach.

For more detail, see Section IV of the Standard (discussed in Chapter 6).

Witnessed audits

CBs must have Quality Systems, Manuals and procedures in the same way as you have and must have a regime of internal audits themselves. Part of this self-monitoring involves witnessed audits.

CBs must witness their auditors from time to time according to a set schedule and may request that you allow someone to witness your audit. This is a burden for the CB of course, especially because there is sometimes resistance to this on the part of companies who fear that they might get a tougher audit somehow. Of course, this should not be the case at all. I do much witnessing myself and can give assurances about this. It makes no difference, honestly.

Furthermore, we all have to be witnessed by our ABs and these can be really difficult to arrange with companies. You are the client and the CB does not want to upset you, but there is an obligation on your part to help the system. Note that Section IV of the Standard under Paragraph 3.1 states that sites are obliged to permit witnessed audits as part of the conditions for certification. I add my personal appeal to that!

BRC competence monitoring

The BRC also has a CB performance monitoring system to ensure consistency. Since 2009, the BRC has been monitoring the performance of all CBs in terms of the following aspects:

- Audit report quality
- Compliance with the Protocol
- Auditor registration
- Uploading of reports in time to the directory
- Commitment and communication

The BRC takes some of this information from the directory (e.g. the report upload time) and some from general performance (e.g. communication) and awards numerical scores accordingly. The scores are then accorded colour coding, thus red, amber or green for each of the five categories. In addition, the scores will lead to an overall colour grading. So it is possible to get an amber or red in an individual category; it is also possible to get amber or red overall. The BRC expectation is that CBs will get an overall green rating. However, should a CB get a red score for an individual category, they must produce an action plan to rectify the situation. Should they fail to do this satisfactorily, they will be given an overall red score. The consequences of an overall red score are that the CB is suspended until they can demonstrate improvements. Two consecutive overall reds will mean suspension for 6 months.

This is happening now and several CBs were suspended in 2010–2011 (note that suspensions can and have also occurred due to specific incidents). The BRC means business here, and poor performance by CBs is no longer tolerated. In a further development, from July 2011, the scores are now published not as colours but as star ratings on the public part of the BRC directory. The aforementioned five aspects can each qualify for a star; thus, a company with four stars means they have a green rating in four of these categories.

Note that the BRC can only judge the CB performance on areas that it is aware of; thus 'Commitment and communication' means with the BRC and other CBs, not between the CBs and their clients.

Referrals

In addition, if a member of the BRC (e.g. one of the major UK retailers) is unhappy about the performance of a CB, they may make a referral to the BRC on an issue of inconsistency and this will be investigated by the BRC. A typical scenario is where a retailer audits a supplier themselves and disagrees with the findings of the BRC audit; perhaps they consider something has been missed or not challenged fully.

Summary

In a world where ultimately it is human beings on the ground who have to operate the system, I do not think that any more could have been done to ensure consistency, and I hope this will give you some reassurance. You might also reflect on the fact that the CBs are under a lot of scrutiny themselves, so they know how you feel when you are being audited.

4 Familiarity with the Standard: Part 1 – Structure and Concepts

It is absolutely vital that you are familiar with the Global Standard before you consider being audited. Note that Clause 1.1.7 requires that you have an original copy available. Knowledge of the principles and format of the Standard is essential so that all the required elements are incorporated into your systems. Understanding the BRC Protocol (Section III) is just as important as knowing the Requirements. It explains what is required before the audit, during the audit and, just as importantly, how to respond to any nonconformities identified. We cover all aspects in this book, of course. Here, we will look at what you need to know before the audit.

> **Key points in this chapter**
> Structure of the Standard
> Fundamental Requirements
> Recurring Themes

Why Choose the Global Standard for Food Safety?

Why join this club? Of course, you have a choice, don't you? You may well have been instructed to achieve the Global Standard by your customers, in which case you do not have a choice. Otherwise, it is not the only food safety standard out there. There are some 'competitors', so you might have other options. I would say this: as well as being recognised by the GFSI (see Section 'Global Food Safety Initiative'), the Global Standard has a genuine heritage and compares favourably with all of them in terms of its comprehensiveness and its accessibility. The Global Standard has been honed over the years by a group that cares about food safety and quality and at the same time has a genuine practical approach to what is possible, which brings us back to the reason that so many of your customers require this Standard.

The BRC Global Standard for Food Safety: A Guide to a Successful Audit, Second Edition. Ron Kill
© 2012 John Wiley & Sons, Ltd. Published 2012 by John Wiley & Sons, Ltd.

Global Food Safety Initiative

The GFSI is a non-profit-making organisation created under Belgian law in 2000. Its stated function is to benchmark food safety schemes such as the Global Standard for Food Safety and, in doing so, reduce the number of audits globally by obtaining agreement from major retailers to accept GFSI-benchmarked schemes as being equivalent. I will not say any more about the organisation here as you can read it for yourself at www.mygfsi.com.

The Standard has been revised five times since the first issue in 1998. There is a set timescale for review of the Standard to meet the Requirements of the GFSI: all recognised Standards must be reviewed every 3 years. For that reason, at the time of writing, we are up to Issue 6 of this Standard. It now has genuine heritage and is well recognised throughout the world.

Issue 6

Issue 6 was published in July 2011. There have been significant changes from the previous issue and the changes are detailed in Chapter 1 for easy reference.

The Structure of the Standard

The Standard is in four sections plus Appendices as follows:

Section I: Introduction and Food Safety Management System
Section II: Requirements
Section III: Audit Protocol
Section IV: Operation and Governance of the Scheme
Appendices

Section I

Section 1 itself is divided into two parts:

The Introduction

The Introduction gives you the history and rationale for the Standard. It tells you how the Standard is managed and reviewed and who has input into it. The key changes for Issue 6 are covered and the benefits of the Standard are described. The terms 'accreditation' and 'certification' are usefully defined.

Note: The process of successfully meeting this Standard is 'Certification', not 'Accreditation'.

There are two other key points to bring to your attention. Firstly, the scope of the Global Standard must be understood. It is a standard for the *manufacture* of processed foods. This means that it can only be used where manufacturing is taking place. 'Manufacturing' can mean anything from a simple cutting or re-packing operation to the manufacture of a complex product from a number of raw materials. Companies who do not change the product in any way, such as wholesalers or storage and logistics companies, may not apply for certification under this Standard (however, they may use the BRC Global Standard – Storage and Distribution).

Secondly, the Introduction refers to Legislative Requirements. The Requirements underline the shared responsibility between manufacturers and retailers for the safety and legality of the products, for example, in detailed specifications. The necessity to meet legal requirements is taken as a 'given', and you should be aware that, overriding all other considerations, the Auditor will be aware of legal requirements for your products.

Food safety management system

Under 2.1, this section introduces two key components on which the Standard is based:

- Senior Management Commitment
- A HACCP-based food safety system

There is an additional paragraph under 2.2 where the format of the Standard is introduced as four main areas where compliance is required and where they feature in the Standard. Table 4.1 sets this out.

Section II: The Requirements

The Requirements of the Standard are outlined here and dealt with in greater detail in Part 2. They form seven main parts or clauses of the Standard as detailed in Table 4.2.

Statements of Intent

Each main subject begins with a SOI as a heading statement above sets of detailed clauses. You can pick them out because they are highlighted. Each one is in effect a mission statement or objective; they set out general principles and objectives and lay

Table 4.1   Format of the Standard.

Senior Management Commitment	Section II Part 1
A HACCP plan	Section II Part 2
A quality management system	Section II Part 3
Prerequisite programmes	Section II Parts 4–7

Table 4.2 Distribution of Statements of Intent and Requirements.

Main clause	No. of Statements of Intent (fundamentals)	No. of other Requirements	Total
(1) Senior Management Commitment and continual improvement	2 (1)	12	14
(2) The food safety plan – HACCP	1 (1)	19	20
(3) Food safety and quality management system	14 (3)	37	51
(4) Site standards	15 (2)	109	124
(5) Product control	6 (1)	30	36
(6) Process control	3 (1)	13	16
(7) Personnel	4 (1)	19	23
Total	45 (10)	239	284

the ground for the more specific requirements that follow. The clauses that follow in turn set out how the SOI is to be fulfilled. Additionally, ten of the SOIs have been rated as 'Fundamental Requirements'. Don't worry, we will explain what these are for soon.

To help you get an idea of scale and an overview of the distribution of clauses and Statements of Intent, see Table 4.2.

There you have it. All you have to do to achieve a BRC Global Standard for Food Safety certificate is to meet 284 Requirements. Yet major nonconformities against only three of them will result in being uncertificated.

The message of this chapter is to treat each Requirement as a separate entity and go through the Standard looking carefully at each one in detail. You must ensure that you have looked at each clause in a positive way and asked yourself how you meet it and what evidence to provide the Auditor with to satisfy them that you meet it.

Part 2 will help you to understand what the Auditor is seeking to satisfy each clause, but you should be very familiar with each one and understand the occasionally subtle differences between clauses in the same section or sub-section.

Note: Issue 5 of the Standard had 325 Requirements in total, so Issue 6 appears to constitute a reduction of 41 or 13%. However, the true picture is that while 30 old clauses appear to have been removed, much of that material has been merged with existing clauses. There are also almost 50 new clauses. So, in effect, the amount of material in the Requirements section has grown, not reduced.

The Fundamental Requirements

The concept of Fundamental Requirements first appeared in Issue 4 of the Standard. They are SOIs that *must be met on the day of the audit*. Thus, there is no allowance

Table 4.3   The Fundamental Requirements.

Clause 1.1	Senior Management Commitment and continual improvement
Clause 2	The food safety plan – HACCP
Clause 3.4	Internal audit
Clause 3.7	Corrective action
Clause 3.9	Traceability
Clause 4.3	Layout, product flow and segregation
Clause 4.11	Housekeeping and hygiene
Clause 5.2	Management of allergens
Clause 6.1	Control of operations
Clause 7.1	Training

for the correcting of either critical or major nonconformities of these issues. There are ten of these as shown in Table 4.3.

It is worth noting that there is a slight concentration of them in Section 3, where systems and documentation must be in place.

If you fail to meet even one of these requirements (i.e. you are given a critical or major nonconformity against one of them), you cannot be awarded a certificate as part of that audit. You will be uncertificated and 'No Grade' will appear on your report, although not, of course, on a certificate. In order to be able to achieve a certificate in future, you would have to apply for a further full audit at a subsequent date.

How they came about

Now that they have been with us for a while, it is worth reminding ourselves why it was decided to designate ten SOIs as Fundamental clauses. Going back to Issue 4 of the Standard, there had been disquiet among the CBs for some time that since the 28-day limit on corrective actions had been introduced, there were certain clauses against which it was very difficult to satisfactorily clear corrective action against nonconformity. It is one of the essential maxims that we need to see a good track record of performance in order to be sure that a company is complying with a requirement.

Example

A company supplying bottled mineral water is not a bad operation overall; they have a reasonable site and good equipment and a small but able management team. Six months before the audit, the Technical Manager left and they struggled to replace the person for a while. When they did get a new person in, they had so much to catch up with that they have completely neglected the internal audit system.

So, on the day of the audit there was a complete failure to meet any of the internal audit requirements. This is a subject that really does require a minimum track record in order to satisfy the Auditor that you have a system and are carrying it through.

In these circumstances before Issue 4 of the Standard we would probably have given major nonconformities for several clauses in that section. The company would then have had 28 days in order to correct these issues, and because they were 'majors', they had to be closed out fully with supporting evidence. So they would do everything possible within that timescale, they set up an audit schedule, they get in an outside consultant to carry out a couple of audits and they send all this in to the CB with all the evidence that they can muster. But truly, the CB cannot say this is a working system from such evidence.

Even if all systems are in place, without a reasonable track record the Auditor cannot really judge whether the company will maintain a good level of, say, internal auditing or whether they have done just enough to get them through the audit. This holds true for all the Fundamental Requirements.

Today, this company would fail a Fundamental Requirement and be told that they will not be awarded a certificate based upon this audit. The only way forward for them is to address the issues and reapply for a new audit when they are ready. Hence, in this example, under the current system at the time, the company was given Grade D (now 'No Grade').

Taking another example, HACCP is literally fundamental to food safety and needs sufficient time for consideration of risk assessment, assessment of CCPs and good management of the system including auditing and so on. It is unlikely that this can be done convincingly in 28 days; hence this would be a failure to meet Clause 2 of the Standard and the site would be uncertificated.

You might think that hygiene issues could be rectified in a 28-day period. However, there was a strong feeling that a company really should be in good order all the time for the sake of food safety. Thus, while it is conceivable that a company could put hygiene problems right in 28 days, it is hard to forgive one that has a dirty plant that goes beyond normal working debris and that has allowed dirt to build up for some time.

Example

A tomato cannery had very good systems on paper, but frankly, on the day of the audit, the site hygiene was very poor. This was more than just the normal working debris of the day; it clearly had not had a proper clean in weeks. There was dried-on tomato debris everywhere, some of it supporting mould growth.

Now, in this case, it might be argued that the management could stop production and make the place as shiny as a new pin in a couple of weeks. The problem here is that the CB can have no confidence in the continued application of the hygiene policies on the basis of a single massive clean-up and therefore it goes down as a failure to comply with Requirement 4.11.

Hence, the company was given a Grade D (now 'No Grade').

The CBs had long been aware that certain nonconformities could not be corrected completely within 28 days even with the best will in the world. At the same time, the clauses they represented were related to important features that any good company must already have in place. Therefore, it was agreed that such requirements must be in place and to the satisfaction of the Auditor with sufficient track record on the day of the audit.

It should be added that a minor nonconformity against a Fundamental clause will not result in being uncertificated but is treated like any other nonconformity.

What you need to do?

No company should rush into applying for an audit against the Standard without firstly ensuring that they can meet all the Fundamental Requirements.

Earlier we looked at making yourself familiar with all the requirements, and this is really essential for these ten. In Part 2, we will look at every single Requirement of the Standard and what the Auditor is looking for, but in general as far as fundamentals are concerned, they need to be well-established systems where you can demonstrate compliance by showing *evidence* of an acceptable *track record*.

Look at the ten clauses in Table 4.3: they are all systems that can be audited internally, including Internal audit itself; so do it. Make them all part of your audit schedule (just remember, an individual must not audit their own activities: ref. Clause 3.4.3!).

This might take time, but that is the proper approach to this. Several of the Fundamental Requirements rely on documentation, so make sure that you have good evidence of compliance for a reasonable timescale.

Colour coding of clauses

You will notice in your copy of Issue 6 that all the clauses are coloured either pale green or a kind of pinky colour I shall call 'peach', or in some cases both. This is intended to be a guide for auditors as to where and how they should audit these particular clauses. Green is intended to indicate that the clause concerns documents, systems and records, so it may be audited away from the production area, while peach indicates that it may be audited during the site tour. There is no hard and fast rule here; the Auditor will decide where best to follow an audit trail. Your Auditor will have been trained in the significance of this and in particular the need to spend 50% of the audit in the factory.

Recurring Themes

We have discussed in Chapter 2 the recurring theme of documented procedures in the Standard. There are some other topics that arise several times in the Requirements, and you should have them as part of your overall culture so that they are not overlooked for specific clauses.

Table 4.4   Clauses that refer to risk assessment.

Clause	Subject
3.5.1.1	Raw materials
3.5.2.1	Raw materials
3.5.4.4	Outsourced processing
4.2.1	Security
4.3.5	High care segregation
4.4.13	High risk air
4.5.1	Water analysis
4.8.1	Changing facilities
4.10.3.1	Metal detection
4.10.6.1	Container cleaning
4.13.1	Pest control
4.14.1	Storage procedures
5.2.3	Allergen control
6.3.2	Calibration

Risk assessment

Risk assessment is a vital tool for you for so many aspects and you should use it as often as possible. Obviously, it is a cornerstone of HACCP, but wherever there are references to risks to product in the Standard, you can use risk assessment to guide you in creating your procedures and to justify them. In Issue 5 of the Standard, there were no less than 27 clauses that specifically referred to risk assessment. However, this number is now lower due to merging of old clauses. Table 4.4 is a list of the clauses that refer to risk assessment aside from HACCP-related clauses, either literally or implicitly. You will need to have evidence of such risk assessments available for audit.

Whether a clause specifically requires it, risk assessment should always be employed in any case when you think that there might be doubt expressed over your decisions on systems and procedures. If the Auditor expresses doubts over your practices, evidence of your risk assessment, if done properly, should be able to justify your position.

Try This

Pick a clause at random from the Table 4.4 and audit your system to answer the following questions:

(1) Did you carry out a risk assessment?
(2) If not, was there a reason?
(3) Was it documented?
(4) Is the result being followed through?

Table 4.5   Clauses that specifically require site plans.

Clause	Subject
2.5.1	HACCP flow diagram
4.3.1	Product risk zones
4.3.2	Access and routes in site
4.4.4	Drains (high care and high risk only)
4.5.2	Water distribution system
4.11.6.2	CIP
4.13.3	Pest control

Do not necessarily limit yourself to the aforementioned clauses; there may be other areas and clauses also where risk assessment is worthwhile.

Site plans

There are now several clauses that require site plans or specific schematics to show different aspects. You may want to depict all the requirements on a single plan or it may be more meaningful to have separate ones. In any case, it is useful to know all the relevant subjects, which are in Table 4.5.

Corrective action

The concept of corrective action is central to most Quality Systems and Standards and naturally appears in the BRC Standard. Clause 3.7 and its sub-clauses are all specifically devoted to the subject. Note that the concept now also includes a requirement to identify root cause of any issue (see Section 'Root cause'). In addition, Table 4.6 shows which clauses also have a specific reference to corrective action.

It should become second nature to build the concept of corrective action into all these systems. Corrective action must be:

- clearly recorded;
- clearly allocated to specific personnel;
- completed within agreed timescales; and
- signed off by the person taking action and authorised by senior personnel.

Table 4.6   Clauses that specifically require corrective action.

Clause	Subject
1.1.3	Management review
2.11.1	HACCP
3.4.3	Internal audit
4.10.3.6	Foreign body detection
4.13.7	Pest control

Root cause

This concept appears at several points in the Standard. Essentially, wherever there are corrective actions involved, you must now also investigate the root cause of the issue and correct this. The requirement for root cause analysis is there for all nonconformities raised at your audit. In this case, you must submit an action plan for dealing with the root cause along with your immediate corrective action (see Chapter 19 for more detail on this). At the subsequent audit, you are required to demonstrate that these have been actioned (see Clause 1.1.10).

The requirement for root cause analysis also appears in the specific Clauses 3.7.1 (Corrective Action) and 3.10.1 (Complaint Handling). So, how do you do root cause analysis? If you search the Internet for 'root cause analysis' using a well-known search engine, you will come up with over 5 million hits; so there is plenty of information out there for you. The simplest is possibly the 5 Whys technique, but you should use one that suits you and your organisation. It is important that you formalise this process and have a technique in house. The auditor might want to see evidence that you have conducted root cause analysis properly. Their training now includes reference to this subject and the various techniques; so they may ask for evidence of the technique you have used.

Review

The concept of review is another that appears several times in the Requirements and is one that must be part of your Quality System management. It is a fact of life that change is necessary and that it sometimes happens in uncontrolled ways anyway. Reviewing your systems is vital to ensure that necessary changes are put in place and that those that have been introduced are controlled. Clause 1.1.3 is concerned with Management Review. This is a crucially important part of your system and is dealt with in detail in Chapter 11. Table 4.7 shows which other clauses specifically require reviews.

Table 4.7 Clauses that specifically require review.

Clause	Subject
2.8.1	HACCP
2.12	HACCP
2.14	HACCP
3.6.5	Specifications
3.11.3	Withdrawal and recall
4.2.1	Security
4.9.3.4.3	Glass breakage
4.13.8	Pest control
5.2.1	Allergen control
5.5.1.2	Test results
7.1.3	Effectiveness of training
7.1.5	Staff competence

Table 4.8 Clauses that specifically require training.

Clause	Subject
3.4.2	Internal audit
3.10.1	Complaint handling
4.2.2	Security
4.9.1.1	Chemical control
4.11.3	Cleaning
4.13.1	Pest control
4.13.2	Pest control
5.2.9	Allergens
5.5.2.4	Laboratory staff

It is so important to review your systems and procedures that you should not necessarily limit review to the subjects mentioned in Management Review and the clauses mentioned in the previous text. For example, Clause 4.13.9 under Pest Control requires you to analyse trends appearing in your pest control results. This too could lead to a review of your pest control requirements.

Note also that you will need to show the Auditor evidence of review for all these clauses.

Training

Training is vital to any company and naturally enough is the subject of a Fundamental clause in the Standard (Clause 7.1). However, there are a number of specific clauses that require specific training throughout the Standard in addition to Clause 7.1. You should be aware of these as a theme and they are listed in Table 4.8.

Section III: The Audit Protocol

When you lay your hands on your newly minted copy of Issue 6, please do not ignore the Protocol section and focus only on the Requirements part. The Protocol is central to the whole Standard, and while you may think that as a company you need not concern yourselves with the detail in there, it contains stuff that you need really to know and do. For that reason, most of Chapters 5, 7, 18, 19 and 20 are dedicated to the Protocol and the Appendices.

Section IV: The Appendices and the Glossary

Section IV of the Standard is a description of the Management and Governance of the Scheme.

There are eight Appendices and, where relevant, they are referred to throughout this book. The Glossary is also a very useful section. All these sections are dealt with in detail in Chapter 6.

The BRC Guideline Series and Other Support Information

For certain subjects, the BRC have published Guidelines on good practice. These have been written since it was felt that illustrations of good practice in these subjects would be helpful to those seeking certification. At the time of writing, Guidelines on the following subjects are available:

Complaint handling
Internal audits
Traceability
Pest control
Foreign body detection
Category 5 fresh produce

More subjects may well be added. I refer to them at relevant points in Part 2, and it is recommended that you read those that are relevant to you. They are available to download at the BRC website (www.brcglobalstandards.com). There is a charge for them.

For the previous issue of the Standard, an Interpretation Guideline for the whole document was available and it is likely that it will be for Issue 6. Keep an eye on the website for that.

In addition, the BRC have published a separate and detailed guide on Product Recall systems, which is also available from their website (BRC Product Recall Guidelines – Product Recall Issue 2, 2007).

Summary

The Standard has several concepts running through it that are the culmination of 14 years of development and evolution so far. Tuning into these ideas will help you achieve certification. Like any evolving entity, I am sure there will be more change in future but I believe that the basic concepts will remain in place.

Quiz No. 2

(1) How many clauses require risk assessment?
(2) What does a green colour-coded clause signify?
(3) When must you demonstrate root cause analysis?
(4) How many clauses require a site plan or schematic?

5 Familiarity with the Standard: Part 2 – The Protocol

Take your copy of Issue 6 the Standard, open it at pages 56–57 and lay it flat on a desk. What do you notice? That's right, the Protocol, Section IV and Appendices now make up more than half of the document. The Protocol describes the process of certification and beyond and includes many issues that are the responsibility of the company (as well as all the other participants). Therefore, it is very important that you are familiar with *all of it*. There are paragraphs in it that you must particularly understand and attend to in order for your audit to proceed smoothly. It is not my intention here to repeat each word of the Protocol, you should read it for yourself, but I am going to emphasise the vital aspects in your audit process and some important issues that can be overlooked by manufacturers. It is wordy and there is much to get through, but help is at hand. What follows is a description of the parts of the Protocol that you need to know before your audit takes place. Some of the elements are discussed in detail in Chapter 7, so are touched on only briefly here. Later, we will look at the parts that happen during your audit (Chapter 10). Then, after we have looked at the Requirements, we will look at the parts that you need to know after your audit has happened (see Chapters 18, 19 and 20).

The audit Protocol is now divided into two parts: the first entitled 'Part 1 – General Audit Protocol' with 18 paragraphs, the second entitled 'Part 2 – Audit Protocol for Specific Programmes'. The following numbers relate to Part 1 of the Protocol unless Part 2 is indicated.

Key points in this chapter
Audit options
Unannounced audits
Audit duration
Nonconformities and their grading

Paragraph 1: Introduction

This paragraph introduces the process and refers to a busy but useful flow diagram that illustrates the certification system from start to finish. One point to note here is that

The BRC Global Standard for Food Safety: A Guide to a Successful Audit, Second Edition. Ron Kill
© 2012 John Wiley & Sons, Ltd. Published 2012 by John Wiley & Sons, Ltd.

it may be necessary for aspects of the Protocol to be updated from time to time. As always in the food industry, issues can arise that are sometimes unforeseen and require new interpretation. For that reason, you should consult the website occasionally (www.brcglobalstandards.com) for news and updates.

Paragraph 2: Self-Assessment of Compliance with the Standard

At this point, you are encouraged to carry out your own pre-assessment before you have your audit. For this, you should have read and understood the Standard to begin with of course. Naturally, there is an expectation that you will purchase a copy and in fact you must have a current, *original* copy of the Standard available (ref. Clause 1.1.7 of the Requirements).

Note: The BRC have published a Self Assessment tool, available free of charge on their website, which consists mainly of a detailed checklist containing all the requirements (BRC Self Assessment Tool, 2011).

You should also be carrying out internal audits as a routine anyway that should assess all your prerequisites and any other systems relating to your scope and the Standard (see also Chapters 2 and 13, Clause 3.4). Another option for you is of having a pre-audit carried out by an outside body. In the past, a number of companies have gone for a pre-audit before their first true audit and it usually consists of a gap analysis against the Standard. We discuss the idea of pre-audits in Chapter 7; however, it is possible that the Enrolment Programme idea may now prove to be a more attractive option for companies (see Section 'Paragraph 3.1: Enrolment Programme').

Paragraph 3: Selection of an Audit Option

This paragraph sets out the various choices open to you. In the Protocol they are introduced here, then detailed in Part 2 of the Protocol. I am going to deal with them here.

If this is your first audit, you now enter the Enrolment Programme. Let's take a look at that first.

Paragraph 3.1: Enrolment Programme

This is a new idea for Issue 6 of the Standard, the idea being to encourage companies who are new to the scheme to have a go and therefore indicate their intentions to customers even if they do not yet meet the Requirements sufficiently to gain certification. Because this is also the entry point for all new companies, it replaces the idea of an initial audit.

What happens is that the company would have a full audit by a qualified Auditor. Nonconformities will be classed as critical, major or minor as usual. However, there is no absolute requirement to close out the nonconformities in 28 days unless the company is seeking a certificate from the first audit. In that case, the normal closing-out sequence is followed (see Chapters 18 and 19) and you will be given a grade as appropriate. However, the company may choose to remain in the continual development phase. In this case, you need only to submit a corrective action plan rather than corrective actions. You would also be required to submit a root cause plan. Instead of receiving a grade such as A, B or C, you would receive a score. The score will be based on the number and nature of nonconformities, and there will be a calculation using a weighting system that will depend on the clauses involved. At the time of writing, the scoring system had not been finalised but it will be spreadsheet-based. The score will be presented on a scorecard that the CB will upload to the BRC Directory along with the report.

The score will appear on the BRC Directory but on the private part only, that is it will not be visible to the public, only to you and any nominated customers that you give access. The score can therefore be seen by your chosen customers, instead of a grade, and your progress towards certification can be monitored. The public part of the Directory is reserved for companies who are certificated. If it takes some time to achieve certification, at least customers can see that the company is moving in the right direction (hopefully) by their scores improving. There is no overall time limit on how long you may remain in the continual development phase. Your next audit will take place within 12 months of this initial audit. If you are confident, you can opt for a second audit any time.

Companies may feel that this is a better option than having a pre-audit, which for many years has been a starting point for the less prepared (or more nervous). See Chapter 7.

Note: It has been agreed that in order to smooth the transition from Issue 5 to Issue 6, that where a site certificated for Issue 5 fails to meet the Requirements for Issue 6 at their initial Issue 6 audit, the site will automatically be in the enrolment scheme. The site will therefore have a scored audit and no certificate.

What about 'initial audits'?

The Enrolment Programme effectively does away with the original notion of an initial audit, which in the past had one or two slightly different features. It is the entry point for all into the scheme even if the expectation is to achieve certification first time.

Note: Up to Issue 5 of the Standard for an initial audit there was an allowance of 90 days for permanent corrective action of certain major nonconformities where this was justifiable. This feature has now been removed from Issue 6 because the Enrolment Programme makes it unnecessary.

You can read more detail about the Enrolment Programme in Part 2 of the Protocol, Paragraph 4.

Paragraph 3.2: Announced audit programme

This is the original idea – a programme of announced, scheduled audits. The Standard says at this point that it is available for 'existing certified sites' only. This is because of the introduction of the Enrolment Programme. Once you are in the normal audit scheme, you may then have further choices available to you. They are the unannounced audit options.

Paragraph 3.2: Unannounced audit programme

Why would anyone choose to have an unannounced audit? Haven't you got enough to cope with? The answer to that question is that some factories have welcomed the challenge to their systems that this scheme brings and have come off none the worse for the experience. There was some fear on the part of companies that major customers (e.g. retailers) would start to insist that they go for unannounced audits, but in reality that has not happened.

Unannounced audits were introduced in Issue 5 of the Standard. In truth, at the time of writing, the original scheme has not proved as popular as had been hoped, so for Issue 6, we are into some new territory with significant changes to the scheme in the form of *two different options*.

Note: The detail on unannounced audits has grown from a few short paragraphs in Issue 5 to over four pages in Issue 6.

The decision whether or not to go for unannounced audits remains with the site; in other words, the whole scheme is still optional.

Before we look at the two options, the following rules apply to both:

- It is only open to those with an existing Grade A, B, A+ or B+.
- You get a '+' with your grade, hence A+ (note the '+' was previously a '*').
- The new certificate supersedes the previous one.
- If you get a C+, you drop out of the scheme for your next audit, which will be announced.
- You must join within 3 months of being issued a qualifying certificate. (*Note*: This is a change; it was 28 days from audit date before with Issue 5.)
- You must supply your CB with certain background information when you choose to participate. See the Standard for the full list (Paragraph 7.2). This should not be difficult as most CBs will be asking for this sort of detail anyway before an audit.

Option 1

Essentially, Option 1 is based on the original concept in that you would have a full audit against whole of the Standard. As far as timing is concerned, obviously the point is that it could happen any time. However, it is stated that the audit shall typically take place in the last 4 months of your due date period. (Why? Because the BRC recognises the cost implication here and that, if your period between audits is foreshortened, is too much.) Also, it is now clearly stated that you may nominate up to 15 days when the audit cannot occur for logistical reasons. This must be agreed with the CB at the outset.

There are two extra points that the Auditor will observe for an unannounced audit. Firstly, they will do the GMP audit, which will take place mostly in the production area, on Day 1. (Why? Because it is in the site where your performance is likely to be most variable and therefore more challenged during an unannounced audit.) Secondly, the Auditor should begin the site audit within 30 minutes of arriving, having done the opening meeting with you.

Option 1: Certification

Certification for this option is fairly straightforward. Whenever your audit takes place, the certificate will be dated to expire as if for an announced audit, that is the anniversary of your very first audit. However, should you withdraw from the scheme and revert to an announced audit (or get a C+ grade, which would mean the same thing), your dates will be reset from the date of your last unannounced audit.

The certificate will also indicate that you are in the unannounced audit Option 1 scheme.

Leaving the unannounced scheme

Once you are certificated to Option 1 Unannounced, you may leave the scheme by transferring either to Option 2 Unannounced or to the announced scheme.

Option 2

Now, for something different. Option 2 is a two-part audit. The first part is unannounced and will include the parts of the audit that make up GMP. In effect, it will take place mostly in the factory, although there may very well be aspects of the documentation done as well. In terms of timing, this Part 1 audit will be done in a 4-month window before your audit due date between months 6 and 10 of your 12-month cycle. There is a reason for this as described in the next paragraph. On the day, the same timing applies as for Option 1, that is the auditor should get the opening meeting done and get into the factory within 30 minutes. As we have said earlier, the focus of the audit under Issue 6 has shifted more to the factory anyway, so the new structure facilitates this.

The big difference here though is that Part 2 of this audit is *announced*. Whatever the timing of the unannounced part, the next stage will happen in a 28-day window before your audit due date, just like a normal announced audit. It will include a review of the site facilities though. It will also include verification of corrective actions from Part 1. This is the reason for the 4-month window for the unannounced part being earlier than in the full unannounced option – to make way for the announced part. Table 5.1 sets out the timescales as a comparison between the two options.

Option 2: Nonconformities, corrective action and reporting

For Part 1 of this audit, nonconformities are dealt with in the same way as a conventional audit with the usual 28 days allowed for corrective actions and so on. A short

Table 5.1   Comparison of Option 1 and Option 2 timescales.

Month	1	2	3	4	5	6	7	8	9	10	11	12
Option 1	3 months to notify					Audit in final 4 months						
Option 2	3 months to notify					Part 1 audit months 6–10					Part 2 audit	

report will be issued to you detailing the nonconformities within 2 days of the Part 1 audit. The rest you can probably guess; for example if you get a critical nonconformity during your Part 1 audit, you will fail the audit as a whole: your certificate will be immediately withdrawn (so not to be recommended).

After the Part 2 audit, a normal audit report will be produced. It will include all the information on compliance plus nonconformities and corrective actions from both parts, and, of course, your final grade is the sum of both parts.

Option 2: Certification

As with Option 1, the audit due date and expiry dates on the certificate will be based on the anniversary of the initial audit date. Of course, in this case, the Part 2 audit will occur in the 28-day window before your due date, as with a full announced audit. Table 5.2 summarises all the differences between the two options (for non-seasonal products).

Seasonal products

For seasonal products, you can now also have an unannounced audit and go for either option. This is an important change for such producers. However, the audit has to be carried out during production, so you must give the CB the expected production dates before the season and you are not permitted any nominated days when the audit cannot occur.

An interesting feature is that for an Option 2 unannounced audit, the normal running order is reversed and the document and systems audit must be done first, at least 28 days before the expected start of the season. The GMP part must be done

Table 5.2 Comparison of Option 1 and Option 2 schemes.

Scheme	Option 1	Option 2
Feature	Single unannounced full audit	Two-part audit with first part unannounced GMP; Part 2 systems and documents: announced
Timing of unannounced audit	Months 3–12, typically 9–12	Months 6–10
Permitted days agreed ruled out	15	10
Process	Single full audit to begin with factory audit within 30 min of arrival	GMP audit to start within 30 min of arrival
Re-audit due date	Anniversary of initial audit	Anniversary of initial audit; Part 2 will take place in 28-day window before due date

Table 5.3   Unannounced audits for seasonal products.

Scheme	Option 1	Option 2
Feature	Single unannounced full audit	Two-part audit with first part systems and documents: announced; unannounced GMP becomes Part 2
Timing of unannounced audit	During season	GMP during season; systems and documents audit at least 28 days before expected start of season
Permitted days agreed ruled out	None	None within the season
Process	Single full audit to begin with factory audit within 30 min of arrival	GMP audit to start within 30 min of arrival
Re-audit due date	Anniversary of initial audit, season permitting	Anniversary of initial audit, season permitting

during the season. Table 5.3 summarises the differences between the two options for seasonal products.

Summary of unannounced audits

The idea of offering the two options is to make the scheme more attractive for companies. Certainly, Option 2 should do this. The challenge to your GMP systems being unannounced is logical as it is in the factory where day-to-day variation is more likely to occur, and at least the systems and documentation part is announced as before. Perhaps, having had an Option 2 audit, you may then feel like the complete challenge of Option 1.

So, if you are up for a challenge and want to show the world the confidence that you have in your systems, go for it!

Paragraph 4: Selection of a CB

This aspect of the Protocol, where the responsibilities of the company in selecting a suitable CB with a suitable Auditor are emphasised, is covered in detail in Section 'Choosing a CB' of Chapter 7. The points raised include the acceptability of the CB to your customers.

Paragraph 5: Company/CB Contractual Arrangements

The CB must agree a contract with you before the audit takes place. This is likely to be in the traditional small print, but it is important that you read it thoroughly as it shall include some of the issues from the Protocol. Payment of the required registration fee is also laid out here. This is discussed in detail in Section 'What will all this cost?' of Chapter 7.

Paragraph 6: Scope of Audit

The scope of the audit is an important subject. The Protocol here looks at how to define the scope, the effect of having more than one location involved in production, exclusions from scope and extensions to scope. We deal with the practical details of all this in Chapter 7, relating to Paragraphs 6.1 Defining the audit scope; and 6.3 Exclusions from scope. Regarding Paragraph 6.4 Extensions to scope, we discuss this aspect in Chapter 20. Regarding Paragraph 6.2 Additional locations and head office audits, as this links in with Appendix 7 of the Standard, we discuss this in Chapter 6.

Concerning the selection of auditors (Paragraph 6.5), it is the responsibility of you, the company, to make sure you convey the correct information to your chosen CB so that they may allocate your audit to an appropriate auditor, that is one who is competent in your field.

What About 'Factored Goods'?

For the past two issues of the Standard, it was permitted to include products that you were trading, but not manufacturing, in your audit scope. You cannot now do this. For some companies who wish to include such items in an audit, they might have to consider going for the Global Standard for Storage and Distribution in addition to this Standard (BRC Global Standard for Storage and Distribution, 2010).

Paragraph 7.1: Audit Planning

Obviously, most of this book is concerned with this subject. There are some important new features here. When planning an audit date, you must take account of how much work you will need to do beforehand. For companies entering the scheme for the first time, if you wish to demonstrate full compliance, then it is unlikely that less than 3 months preparation will be adequate. For all audits you must ensure that the Auditor sees production of as wide a part of your scope as possible. That is your responsibility. As a consequence, if you have radically different processes going on at different times of year, more than one audit may be necessary. For example, if you are canning fruit during a short season and in another part of the year you do something completely different that requires different processes, such as dried pulses, then a separate audit is likely.

Paragraph 7.2: Information to be Provided to the CB for Audit Preparation

This is new for Issue 6 in the sense that you would always have been asked to supply certain information to the CB in the past but the Protocol now presents you with a formal list of items that the CB is likely to ask for. There are eight items listed that the CB 'may include' in its request for information and I think it likely that these items will always be requested. Some of them are listed because they help the auditor with their preparation for the audit; some of the information will also help the CB estimate the audit duration.

You must also have a copy of any previous report available. Why? Because there are occasions when the previous report was done by a different CB.

Paragraph 7.3: Duration of the Audit

From the very first days of Issue 1 of the Standard it was clear that while some sites could be comfortably audited in a day, some were too large and complex for this and some very long days were put in by Auditors. By now, with Issue 6, we are in a completely different situation. The Standard itself has grown over the years, as have the Requirements for auditors to record evidence seen in some detail. To help CBs to work out how long they should spend on each audit, we now have an Audit Calculator. This includes consideration of the following factors: size of site, number of employees and number of HACCP studies. Not only does this help us to calculate the time an audit should take, it also helps to ensure consistency between CBs. As it says in Issue 6, the typical audit time on site is 2 man days, so we have moved a considerable distance from Issue 1.

You can download the Audit Calculator document by going to the website www.brcglobalstandards.com: click on 'Standards', then go to 'Food', then 'Downloads' and you will find it in the list (see also Chapter 8).

The paragraph also mentions time required for review of any corrective evidence and reporting time.

A key point for the manufacturer here is that it is your responsibility to provide the right information again to the CB so that they can assess the likely duration. You should also inform the CB of any shift patterns or schedules, including night-time working.

Note 1: A 2-day audit can be carried out over 1 day by two auditors.
Note 2: A 'man day' = 8 hours.

Paragraph 8: The On-Site Audit

This paragraph gives you details on the format of the audit. This is discussed in detail in Section 'The Format of the Audit' of Chapter 10. Note one important feature: the most senior Operations Manager on site (or deputy) must take part in the opening and closing meetings. This is reflected in Clause 1.1.9 of the Requirements.

Paragraph 9: Nonconformities and Corrective Action

The importance of understanding the whole of this paragraph cannot be overstated. We will look at nonconformities here and leave discussing corrective action until Part 3: Chapter 18.

Paragraph 9.1: Nonconformities

Firstly, Paragraph 9.1 defines critical, major and minor nonconformities.

Critical

I believe that most can agree on what would constitute a critical nonconformity. Where a food safety issue or a transgression of legal requirement is decisive and could result in crisis, the Auditor has no alternative but to give a critical nonconformity.

Major

The definitions of both major and minor nonconformities are equally clear, but it must be said that there is sometimes confusion with companies as to why a major or a minor has been given. Concerning major nonconformities, for me, it is easier to begin by saying what they are not. A nonconformity is not a major because of the perceived importance of the clause in question. It might be thought that some clauses have a more immediate impact on food safety than others and therefore any nonconformity against such a clause must be major. In fact, the Auditor will judge the degree of conformity for each clause on its own merit. Thus, the definition in the Protocol for a major as 'a substantial failure to meet any Statement of Intent or clause' should be understood. Alternatively, any situation that gives the Auditor doubt as to product conformity will also lead to a major.

In addition, a major nonconformity may be given where there is significant doubt as to the conformity of the product.

Examples

Clause 4.7.1 requires a system of planned maintenance. At a bakery, the Auditor found that there was no system at all in place for planned maintenance: all maintenance jobs were reactive.

Because they completely failed to meet the main element of the clause, a major nonconformity was given.

Clause 3.9.2 is a straightforward clause that requires a test of traceability to be carried out. At a manufacturer of canned peaches, the Auditor found that there was no test of traceability at all.

Thus, again, the company had completely failed to meet the requirement and a major nonconformity was given.

In the aforementioned cases, the Auditor had a simple decision to make because the requirements had not been met at all. In other cases, a major will be given because the company has substantially failed to meet a clause such that food safety could be affected.

> ## Example
>
> A high-risk bakery was found by the Auditor to have numerous damaged baking trays across the site. A bandsaw was damaged with missing pieces of metal and blades on sauce cookers were worn. The Auditor considered that there was a danger of metal contamination and gave the supplier a major nonconformity under (for Issue 6) Clause 4.9.2: Metal Control.

Minor

A minor nonconformity differs from a major in that it is one where conformity with the clause or Statement of Intent is not fully met but products may still be considered safe. In other words, if the Auditor considers that you have in fact partially met the clause and there is no risk to product conformity, a minor will be given.

In Table 5.4, the examples of minor nonconformities are all from the same site. It should be seen that in these examples a part of a system has complied but not all. For example, the company were carrying out a traceability exercise, so were meeting some of this clause; however, they had neglected to include packaging. Hence, a minor nonconformity is given. Again, one door does not meet proofing requirements, all the rest were in good order, and hence a minor nonconformity is given, and so on.

Figure 5.1 is a schematic that illustrates the decision process for grading nonconformities. At the time of writing, this is used in the current training courses for Auditors and is reproduced with the kind permission of the BRC.

Table 5.4 Examples of minor nonconformities.

Nonconformity number	Detail of nonconformity
1	The company was not tracing packaging in their traceability test exercise.
2	Ceiling girders in a store room had a build-up of dust and cobwebs.
3	Door open in Unit 4, allowing ingress. Plastic strip curtains not giving complete proofing.
4	Locker segregation was not being enforced.
5	Glass breakage procedure lacked detail and accountability.
6	There was no logging of fruit temperature during defrosting.
7	There was no mention of a policy forbidding false nails in the staff rules.

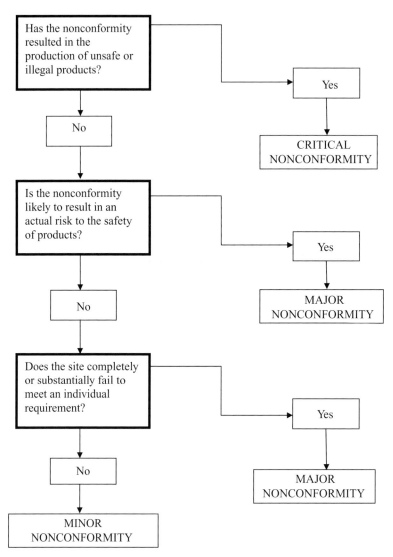

Figure 5.1 Decision tree to determine nonconformity grading. Reproduced with the kind permission of the BRC.

Try This

Determine whether the following nonconformities are critical, major or minor. (*Clue*: There is only one critical.)

(1) A company had colour-coded its food-handling scoops to distinguish between raw materials containing allergens (red) and those that were allergen-free (green). The Auditor found one white scoop being used for allergen-free product.

(2) A company with a high-risk cooking area did not have full segregation between that and the mixing area that was considered to be low risk.
(3) The company packed retail packs to a nominated net weight, but on line this was done by visual volume when packs appear to be full and there was no formal weight control or records kept.
(4) The company had delayed their second audit for 1 month because they had to recruit a new Technical Manager.
(5) A member of staff on the packing line was wearing a watch.
(6) Twenty cases of finished product were noted directly on the floor of the finished goods store.
(7) The company had regular pest control inspections but no in-depth surveys were carried out.
(8) The company had carried out a traceability test but it was 15 months after the previous test.

Answers are given in Appendix 1.

As described in Chapter 1, one of the significant changes made for Issue 6 was that a number of clauses have been rolled up into larger single clauses. Apart from appearing to lessen the overall clause number, this also has the effect of making it more rational to give a minor nonconformance against certain issues.

There are two related aspects. If the Auditor is in a situation where several minor nonconformities have been raised against the same clause, there comes a point when they should consider raising a single major nonconformity. There is no hard and fast rule about how many minors make a major because it will depend on the circumstances. But it will be well before the Auditor gets to double figures!

Furthermore, the Auditor must not 'cluster' a significant number of minor nonconformities together as a single nonconformity. At a certain point the Auditor must decide whether to award a major nonconformity.

Summary

In this chapter, we looked at the first part of the Protocol to include all the aspects that you need to know before the audit takes place, including your options for audit. I have also included information on nonconformity levels because I think it is something you need to know about before the audit. In particular, be aware of the part you must play in co-operating with the process and giving proper information to your CB.

There is much more to the Protocol, of course, but the rest relates to events after your audit, so we will return to it later (Part 3).

Quiz No. 3

(1) Identify which clauses apply for each of the seven nonconformities in Table 5.4.
(2) How many days are you allowed to block out if you go for an Option 2 Unannounced audit?
(3) Which clauses apply in the 'Try This' exercise?

6 Familiarity with the Standard: Part 3 – Section IV and the Appendices

The Standard is very concise; there is little superfluous material in it. Just as with the Protocol, the Appendices have information that you need to be aware of. Indeed, they are much expanded and there are now ten Appendices in the Standard. In this chapter, we will go through them all and have a look at Section IV; however, we will focus specifically on two of the Appendices, numbers 2 and 7 here, because much of the content of the others is dealt with elsewhere in this book. Appendix 2 is of particular significance concerning production zones. Appendix 7 has some interesting concepts concerning multiple site audits.

> **Key points in this chapter**
> Risk zones
> High care and high risk
> Multiple sites
> Scope extension

Section IV: Management and Governance of the Scheme

This section is completely different from Issue 5 of the Standard where it was devoted to a description of the BRC Directory. In Issue 6, there is new information about the Management and Governance of the Scheme. Some of the subjects have been looked at in Chapter 3.

The section begins with a brief outline of the Requirements for CBs and refers to the document 'Requirements for Organisations Offering Certification against the Criteria of the BRC Global Standards' that we referred to in Chapter 3. There is also guidance as to how to locate CBs that we discuss in Chapter 7. The section looks at the Governance and Strategy Committee, then the TAC, Co-operation Groups and consistency matters, all of which were covered in Chapter 3.

The Appendices

The Appendices' titles are briefly laid out in Table 6.1.

The BRC Global Standard for Food Safety: A Guide to a Successful Audit, Second Edition. Ron Kill
© 2012 John Wiley & Sons, Ltd. Published 2012 by John Wiley & Sons, Ltd.

Table 6.1   The Appendices.

1	The Global Standard for Food Safety and its relationship with other BRC Global Standards
2	Guidelines on defining product risk zones
3	Qualifications, training and experience requirements for Auditors
4	Product categories
5	Certificate template
6	Certificate validity, audit frequency and planning
7	Audit of multiple sites
8	Extension to scope
9	Glossary
10	Acknowledgements

Appendix 1

This is a brief Appendix to introduce the three sister Standards to the Global Standard for Food Safety. We need not look at this in detail here.

Appendix 2

This is an important new Appendix that details the concept of different production risk zones in a factory. *It is essential that you are familiar with this Appendix.* The four possible zones are:

(1) Enclosed product area
(2) Low risk – open product area
(3) High care – open product area
(4) High risk – open product area

A site may have one or many of these production zones and the idea is to be able to classify these precisely in your site. Much thought was given over to this during the review process and in particular the differences between high-care and high-risk areas. Remember that when we discuss these terms in this context, we refer to the factory zone, not the product (although the Glossary retains a definition of high-risk product as well). It is important to read through this Appendix carefully to establish which zones you have on site.

Enclosed product zones should be self-explanatory. Within your site, you are likely to have some at least, for example your finished product warehouse. Some sites will even have their production areas all enclosed, especially for certain liquid products. 'Open product areas' refers to where the product is exposed.

Note: Open product areas will include products in natural shells such as eggs and nuts.

Low-risk open product areas are also likely to be found in most sites. The Appendix lists four bullet points that define low-risk areas. The product is open but the nature of the product means that growth of pathogens will not be supported; either that or they will be killed off before consumption by cooking anyway. Your main concerns are likely to be foreign body contamination or cross contamination with allergens. However, exercise some caution, because even in products that we store at ambient temperatures, some pathogens may survive even if they do not grow. Chocolate is well known for the potential to support the survival of salmonella. For example, in 2006, there was a major incident when a very well-known chocolate company in the United Kingdom was found to have product that tested positive for salmonella and was linked to an outbreak of the disease. Thus, your low-risk zones as defined here may yet require special care and attention.

For many, the difference between high care and high risk will be of great significance not least because the requirements differ. For example, if you have a high-risk area, you must have a system of air changes for filtered air (see Clause 4.4.13). In Chapter 14, we look at the clauses that relate to high-care and high-risk zones. The Standard has also helpfully emboldened the words *high care* and *high risk* whenever they appear in the clauses.

Note that the definition of high care has four bullet points that must all apply. For high risk, there are also four points that must apply. The most significant difference between the definitions of high care and high risk here is the degree of cooking that the product has undergone before entering the area. For high care, the defining point is that products must have received a process that typically reduces microbiological contamination by 1–2-log reduction, while for high risk, it is a full cook that is defined as typically a 6-log reduction in *Listeria monocytogenes*.

After all these definitions, you will find a very useful decision tree. Note that it gives examples of certain foods at each step – remember that these examples relate to an area where these products are most at risk. Remember also that the decision tree is only a guide.

Appendix 3

This gives a brief resumé of the general qualifications and experience of Auditors. This has not changed with the review for Issue 6 and provides you with an idea of the basic requirements to qualify as an Auditor before, alongside the fact that they have to demonstrate competence in certain product categories (see Chapter 3).

Appendix 4

This table details all the 18 product categories that apply to the Standard. No doubt, there are some specific foodstuffs that do not feature on the lists, but it should be possible to categorise everything that is likely to arise. There are some minor changes to the listings (see Appendix 4 of this book).

Note: The list continues to include pet food (Category 11).

Appendix 5

The certificate template is largely unchanged for Issue 6 but the new audit options are accounted for.

Appendix 6

This table replaces a flow diagram of sorts in Issue 5 and gives various scenarios for due dates for all the audit options. Timescales are discussed in Chapters 18 and 19.

Appendix 7

This is an important new Appendix. It deals with two different situations concerning multiple sites. Note that the subject is also touched upon in Paragraph 6.2 of the Protocol. There is one aspect that appears under Paragraph 6.2 that does not appear in the Appendix, which we shall discuss in the first situation.

(1) Multiple site production

The first is where more than one site is involved in the scope of an audit. This can occur where production or storage are spread over separate buildings. You can think of this in terms of the separate areas or rooms of a factory being separated geographically; for example, a drinks company making the product at one site and carrying out the bottling at another site, or maybe two sites in the same town. The key here is that the individual sites meet the criteria set out in five bullet points in this Appendix.

The sites must all be part of the same company and using the same FSQMS. They must be involved in the same manufacturing process and thus not be supplying customers individually. Finally, there is a limit of 30 miles or 50 km on the geographical spread of the separate units. Why? To allow for a single audit in the allotted timescale.

This is considered to be an exceptional situation here, but there are examples of such production and it is an alternative to having separate audits at each site. One relatively likely situation is that of an off-site storage facility that also meets all the criteria. This is discussed under Paragraph 6.2 of Part 1 of the Protocol. The difference is that the facility is *required* to be part of the audit provided it falls within the 30-mile or 50-km radius unless it is specifically excluded.

Note: A storage facility may not be audited against this Standard in isolation; one would have to use the Global Standard for Storage and Distribution if a 'BRC Audit' was required.

(2) Head offices

The second part of Appendix 7 concerns companies that have a separate head office. Of course, this is usually where there are several sites as part of a group. There might be certain parts of the food safety systems that are created and run at head office, away from the actual production. Commonly, these are areas such as raw material purchasing; it may also be those parts of the FSQMS that are controlled at the centre, such as HACCP, internal audit and so on. For the purposes of the audit, there are two possible approaches that are set out in this Appendix:

The first is where each manufacturing site is audited and all material and evidence of activities carried out by the head office are audited at the site. Thus, the Auditor must be able to have access to all necessary evidence and consequently might need to interview head office personnel. The Appendix sets out the various methods by which this might be done, all involving the modern communications options available to us, ranging from old-style fax to video links. It is made clear that any failure to provide the evidence required may lead to nonconformities being raised. Also, the failure of the required key staff to be available even at head office during the audit might also lead to a nonconformity (ref. Clause 1.1.9).

The second option is to carry out a separate head office audit. This is slightly more complicated, but workable. The rule is that the head office audit must be carried out first. Obviously, it is only a part-audit, so no certification process or formal reporting will happen at this stage. Nonconformities might be raised and these shall be corrected in the normal way. Then, when the first production site is audited, the head office nonconformities will be reported along with any found at the site. Those that have not yet been closed out will count against the site and will be included in the assessment of the grade (or 'No Grade') of the site. There is then the normal 28-day period for the correction of both sets of nonconformities. The formal report issued will refer to the head office audit as well, including addresses and dates.

The likely scenario is that there will be more than one manufacturing site; so when the others are audited, the nonconformities raised at the head office will not be reported unless they have not yet been corrected.

Regarding the certificates issued, these will only be for the manufacturing sites and the re-audit dates will be in the usual cycle. For the head office, a re-audit must occur before the anniversary of the first manufacturing site audit.

Appendix 8

This is another new Appendix that gives further information on extension to scope. The subject is very briefly mentioned in the Protocol under Paragraph 6.4. In this Appendix is guidance on whether an application to extend scope should require an extra site audit. For example, the CB should make a revisit if the extension involves new technology or methods, new production areas, new hazards and so on. Such a revisit may not be a full audit; it will depend on the nature of the extension. Naturally, any nonconformities must be corrected in the normal way within 28 days. There are three significant points about the result of such a revisit:

(1) A full report is not required to be uploaded to the Directory but a short report should be issued.
(2) The certificate shall be reissued with the complete new scope.
(3) The certificate grade does not alter despite the level of nonconformities found at the revisit.

To expand on point 3, should the CB find a situation during an extension audit that calls into doubt the validity of a certificate (a critical nonconformity for instance), then they may withdraw your certificate and require a new full audit at any time (see also Chapter 20).

Appendix 9

This is the Glossary. Once again for Issue 6 it has been expanded and remains very useful.

Appendix 10

Finally, a list of the contributors to the Standard that you may read for interest but need not detain us here.

Summary

As you have seen, there is some very important information contained within the Appendices, which you should be aware of; so do not overlook them in your enthusiasm to digest the requirement clauses. Two of them in particular (2 and 7) are new concepts for Issue 6.

Quiz No. 4

(1) Raw meat portions are being packed into retail packs. What type of production zone is this?
(2) Sandwiches are being made using cooked meat and salad fillings. What type of production zone is this?
(3) How many product categories are there?
(4) A site has a storage facility that is 50 miles away from the manufacturing site. Must it be included in the audit?
(5) What should happen if a critical nonconformity is raised during an extension to scope visit?

7 Final Steps to the Audit

In this chapter, we move on a stage and look at what to do next. This includes choosing a CB and creating the right situation for the audit to happen as smoothly as possible.

Key points in this chapter
Choosing a CB
Agreeing the scope
Exclusions and extensions
Key staff
Records
Pre-audits: Are they worthwhile?
Providing the right facilities

Choosing a Certification Body

For most countries now there is a good choice of accredited CBs out there for you. You might even choose one that is not yet accredited: CBs are permitted to carry out audits while they await their own accreditation. Firstly, they have to register both with their AB and the BRC; secondly, they must attain accreditation within 12 months of registering.

It is often said that the best form of advertising is word of mouth. In this case, food companies may find it difficult to discuss with their peers and competitors who they would recommend to use, so in some respects you are left to your own devices when choosing a CB. This chapter should help.

Where to find them

CBs must register with the BRC and in order to be accepted must meet all the criteria laid down by the BRC (see Chapter 6). All registered CBs are listed on the BRC Directory (www.brcdirectory.com – see Chapter 8) and the list indicates which countries

The BRC Global Standard for Food Safety: A Guide to a Successful Audit, Second Edition. Ron Kill
© 2012 John Wiley & Sons, Ltd. Published 2012 by John Wiley & Sons, Ltd.

they operate in. In the United Kingdom, the UKAS website also lists all those CBs accredited by them (see www.ukas.com) and for each entry there you can also download a copy of their Schedule of Accreditation (see Chapter 6). In other countries, the various websites of the ABs may be helpful but your first reference point is the BRC website. The reason for this is that CBs must be under contract with the BRC to be permitted to operate this scheme (see details of contract in the following text).

Location of the CB

At the time of writing, the number of registered CBs is well over 100 worldwide. One significant factor in your choice will be their geographical location.

Within a country, CBs will generally have Auditors who are not based at their head office but are operating from their home base. The main implications for you will be the availability of an Auditor who is not only competent in your field of work but is reasonably located so that they are likely to be available when you want them and at a reasonable cost.

Note that there is a requirement for CBs to rotate auditors so that they do no more than three audits in a row at any one place. Obviously, there are practical issues with this and there can be justified reasons for staying with the same auditor. However, in practice, it does mean that you will not always get the nearest auditor to your location, where the CB is applying this rule.

'Approved' CBs

At the time of writing the previous edition it was also the case that certain major retailers in the United Kingdom and elsewhere operated their own approval systems of CBs in addition to their accreditation and registration with the BRC. Certainly in the United Kingdom, this factor has diminished greatly it may still be the case in other countries, so you still need to be aware of this possibility and you may need to check that a CB is approved by all of your customers.

Auditor competence

We looked at CB competence in Chapter 3. The CBs must ensure that their Auditors are and remain competent to do their work. Once an Auditor is qualified to do this work in a certain field, this involves a pattern of internal auditing and refresher training. However, it is the company's responsibility to give the CB all the information necessary at the time of agreeing a contract, so that a suitable Auditor is selected for your audit (see Paragraph 7.2 of the Protocol). At the same time, you should ensure that the CB has the auditor competence to cover the scope of your audit.

What will all this cost?

You will no doubt compare fees charged by the CBs. There is no standard fee for a Global Standard Audit and free market forces apply. Some CBs will give you a very itemised fee with separate costs for the audit, the administration, any necessary translations and the issue of a certificate while others have a simple one-off fee. Note that most CBs quote a fee for the audit and that the variable travel costs will be charged separately.

You must also be registered with the BRC as part of your certification process. There is now a fixed fee for registration with the BRC of, at the time of writing, £125. The CBs are under contract to the BRC to charge you for your registration. The fee is collected by the CB and paid on to the BRC subsequently (see Section 'Registering with the BRC'). Note that CBs are very likely to want to receive payment before they issue a certificate.

CB resources

Assuming you have arrived at a shortlist, try asking CBs how many Auditors they have available who are qualified to audit in the field of the food industry in which you operate. You want a CB who will be able to do the work more or less when you want it without incurring too much cost for travel. If they are limited for Auditor availability, they might not be able to fit you in for a long time or if they can it might mean sending someone in from the far end of the country (or continent).

Availability is also important in the long term. It is vital that, assuming you achieve a certificate, the CB has the resources to be able to meet the deadlines imposed for any of your subsequent audits (see Chapter 20).

You might also ask them how long they take on average to complete and send out reports and try to get a feel for the efficiency in terms of administration. There is an overall requirement for the CB in most cases to complete the cycle of work and be able to issue reports and certificates in 49 days. A major factor in being able to achieve this deadline is in fact the efficiency with which you correct any nonconformities within the 28-day deadline – this is one of the aspects of CB Competency monitoring (see Chapter 3). You should ask how well the CB does in meeting these time constraints, however.

The administrative side of this work is very burdensome for the CB and will form a major part of their fee structure. It is very important that you feel that you will get good value here and that they can achieve a good turnaround of work.

The Application Form and Contract

Once you have made your choice, the next thing that happens is that you are asked to sign a contract or application form that will probably have detailed terms and conditions attached. The CBs are required to do this although each one might vary a little.

The application form is likely to ask you for certain details that are important for the CB to be able to assess the likely duration of the audit. They will include: the number of production lines, number of different HACCP studies, number of staff, size of production areas and so on (see Chapter 5, Paragraph 7.3 'Duration of the Audit').

The CB now also needs certain information from you such as a site plan, management structure, HACCP flow diagram and so on. The full list is given in the bullet points to Paragraph 7.2 of the Protocol. All these requests help the CB to assess you a little before the date of audit so it will be beneficial to you to help them as much as you can at this stage.

The questions on the form will also ask you to define the scope of the audit and any exclusions that you propose. The CB must then go through a thorough review process when they have all the information in order to confirm the likely duration, to confirm the allocation of the audit to a particular Auditor and to confirm their acceptance of any exclusions to scope that you have proposed (see Exclusions to the Scope). For these reasons, it is very important for the CB that you provide all this information as soon as possible and give the CB the proper time to make the arrangements for your audit.

Contract

It is important to read the terms and conditions carefully. As Sam Goldwyn is attributed as saying, 'A verbal contract isn't worth the paper it is written on'. I know that this is one of those aspects of modern life that we would all rather pass by but there is a point here. There are likely to be some important clauses in there that you should be aware of. This is not just a standard list of disclaimers but a binding contract, which requires you to do certain things many of that are dictated by the Protocol.

One example is that the CB is likely to ask you to sign your consent to allow them to send a copy of your report to one of the major UK retailers. Despite the fact that the BRC Directory system is there now for your clients to access, it is still the case that many require the CBs to send them your reports directly. This may be a part of the application form itself or as a separate consent form. Another example is that the CB is likely to put a requirement in the terms and conditions that you inform them of any significant changes or product recalls, as required by the Protocol. It is important to be aware of this as failure to do so could potentially result in certificate withdrawal.

Your site and location

You must make clear to the CB any site that is involved in the processing or packaging of your products at any stage of the manufacture. It is possible, of course, for a process to be carried out over more than one site, especially if you include product storage. In such cases the fact that more than one site was involved will be noted both in the report and on the certificate. As far as auditing goes, if a company has off-site storage facilities that they directly manage this will be included in the audit up to a radius of 50 km (see Section 'Agreeing the Scope' and Chapter 6 Section 'Audit of Multiple Sites'). If you are using a sub-contractor to carry out part of your process this must be managed as detailed in Clause 3.5.4 of the Requirements (see Chapter 13).

Agreeing the Scope

The scope of the audit means those products, or ranges of products that you wish to be included in the certification. This will be completed by you on the application form and it is crucial that you answer this fully and that the scope is agreed before any work takes place. Note that the scope is also confirmed in the Opening Meeting on the first day of the audit (see Chapter 10).

The Protocol (Paragraph 6.5) requires that you provide the right information to the CB so that they can ensure that the right auditor is chosen. From your point of view, this is important because you need to be certain that your chosen CB is qualified to audit your operation and can demonstrate competence in your field. From the CB's point of view, this is important when it comes to allocating an Auditor. Also, when issuing a certificate the CB must be very precise about what products are certificated. The scope of the audit should be a precise list of the various product types that you wish to be included.

Example

A small bakery completed an application form and for the scope of the audit entered 'all products supplied by us'.

This clearly will not do. The final scope agreed was: 'manufacture and packing of long life ambient stable cakes, fruit pies, muffins and shortbreads and the case-forming and packing only of filo pastry'.

Things have moved on a little since this example occurred and the BRC have issued guidelines to CBs concerning the wording of scopes. The scope of the audit must be agreed before it takes place and this must be site specific as well as product specific. The scope should describe the process and the nature of the pack. The use of terms such as 'Manufacture, Production' etc may be used only if they resultant products are clear. Meanwhile words like 'Procurement, Receipt, Intake, Storage', which frankly add nothing to the knowledge of the user as they are bound to be part of the process, are not required. Furthermore words such as 'Sales, Marketing, Growing, Distribution' are not permitted in the scope. In short, the scope must describe the products characteristics, which are based in the process and form of pack. Thus:

Scope: Pasteurised Whole, Semi-Skimmed and Skimmed Milk in 1 L and 2 L Plastic Retail Bottles

is an acceptable scope, while:

Scope: Pasteurised Milk

is not. For appropriate products we may need to indicate other food safety features such as being packed under modified atmosphere, or that they are chilled or frozen.

There is good reason for the scope to be as precise as this because if in time a company adds new products or processes to their range which do not come within the agreed scope then they would not be considered to be within the certificated range. If wanted in the scope, the company would have to either include them in the scope at the next audit or apply for an extension to scope audit if they need to add to their scope between scheduled audits. For similar reasons, the idea of exclusions to scope was introduced with Issue 4.

Note: Now that factored goods are no longer permitted in scope, a description of how they are managed by the company may be included in the audit report.

Question: My company handles mostly factored goods with a small amount of cheese cutting and packing. Under Issue 5 we were allowed to have all this under a Food Safety audit. What do we do now?

Answer: To retain certification for all that you do your best option is to have a joint audit of the Food Safety and the Storage and Distribution Global Standards. Many CBs should be able to offer this service.

Question: My company packs food hampers. Some of the items are packed by us, some are brought in. Can we be certificated under the Food Safety Standard?

Answer: As long as at least one of the components is manufactured at the site this can be included in scope. The components not manufactured on site will be treated as if an ingredient. The Requirements of Clauses 3.5 and 3.6 shall be met.

Exclusions to the scope

You are permitted to exclude certain aspects of your company from the scope of a BRC Standard Audit. The kinds of thing you may exclude are products that are produced in a separate area of the factory or products produced on different equipment. However, as it says in the Standard, exclusions from scope are only permitted by exception. If you wish to exclude any products or processes from the audit then this must be positively declared at the outset. They must be agreed before the audit. Any exclusions to the scope are noted and will appear on the final certificate.

The following scope would be permitted:

Scope: Manufacture of canned peeled plum tomatoes
Exclusions from scope: Passata in Glass Jars

The following scope is not OK:

Scope: Unsliced Bread and rolls
Exclusions from scope: Sliced Bread

Why not? Because the sliced bread is made on the same production lines as the unsliced.

The audit must include the entire process: you cannot exclude part of a process either. Thus, the following would not be permitted:

Scope: Baking of unsliced and sliced breads, rolls and cakes
Exclusions from scope: Mixing of doughs

Registering with the BRC

This is an automatic requirement for anyone applying for a BRC audit. This will be taken care of for you by your chosen CB. The CBs themselves are in turn under contract to the BRC to administer this system, which requires that you register with the BRC. You will be given a site code, which will appear on your report and certificate and which will always apply to that address. Should you change address, a new code will be issued. Should you leave the scheme and later re-join, you would retain the code. The BRC then requires the CB to keep them informed of the results of the audit by submitting the final report in electronic form.

Note that confidentiality is assured by this process but you should specify to the CB any information that you give that you may reasonably ask not to be reported on. This too is confirmed in the opening and closing meetings during the audit.

For the Stakeholders, the benefit is that they now have a one-stop place where they can monitor the status of their companies and save considerable administrative time contacting the CBs individually for this information. At the time of writing, this process has not yet found complete favour with some of the Stakeholders and CBs are still obligated to send them copies. However, in theory, for the CBs the benefit is a likewise saving in administration in that they now only have to deliver this information to one place (see Chapter 8).

Fixing a Date

This sounds straightforward but there are some points to think about. For an initial audit you need to consider an appropriate time for the CB to be able to see your production as fully as possible. In the extreme, for seasonal products such as canned salmon, canned fruits or tomatoes this may be for only a few weeks in the year. Even in a less extreme situation the CB will not be satisfied by seeing only half an operation, they must see production happening.

You also need to look at your staffing and make sure that a date is agreed when your key staff are available. Check for any reasons for absence, especially those that might recur annually such as holidays.

Any date that is agreed will go down as your initial audit date and thereafter will be referred back to for all subsequent audits. The timing of subsequent audits will have to take place within a 28-day window before anniversaries of initial dates (saving that a Grade C certificate means a 6-monthly return). Therefore, it is particularly important to avoid periods of low (or even no) production and staff absence, that are likely to repeat.

Duration

The likelihood is that the audit will take up to 2 days (see Chapter 5). Some large or complex sites may take more. The CB should judge this fairly, based upon the information that you give them. They, like you, will be aware of the cost implications while needing to ensure that you are audited properly.

It is important to take the duration into account when arranging a date because again you will need to make key staff available for most or all of the time the Auditor is on site.

Key Staff

Once you have fixed the date, as it approaches you should check again that the key staff are going to be in and available. The personnel that the Auditor is likely to need to see are the Technical Manager and/or Quality Manager, plus senior people in the following areas: Production, Maintenance or Engineering, Hygiene and/or Pest Control (in short, probably a good part of your HACCP team) and Personnel Management. Note that the Standard also requires that your most senior Operations Manager on site (or deputy) shall attend the opening and closing meetings. Obviously, some of these roles may be combined within one position. The Auditor should also pick up on other staff as they tour the site to speak to and may also note some names to follow up later when checking on training systems.

During the opening meeting, the Auditor should set out a programme for the audit and should ask when certain people will be available just in case they cannot be present for all the time. Not everyone in factories keeps 'factory hours' and where there is an office-based function, such as Personnel, this may even be part time. The Auditor should not assume that they will be available late in the afternoon.

Using consultants

If you feel that you do not have sufficient technical expertise in house, it is perfectly proper to use someone from outside with the right experience to guide you through the preparation for the Global Standard process. Choose a consultant with care, ask them about their experience in this field and perhaps their success stories. Also, you must take them on as part of your team and not allow them to work in complete isolation (see also Chapter 10).

Availability of Records and Other Evidence

This is really vital and bears repetition. The Auditor is going to ask for lots of documents from Policies and Procedures to cleaning records, pest control records, training records and all in between. From the Auditor's point of view, there is nothing worse (it bears repeating) really nothing worse than having an audit continually interrupted by prolonged searches for documents.

Long are the hours that Auditors have spent in offices staring at the walls while a lone Technical Manager has gone off on some long expedition to find a piece of paper. Often, they return only to find that they are sent off again for an additional piece of information. Not only does this give a poor impression and will lessen the confidence of the Auditor in your ability to maintain your systems, it also makes the audit take so much longer (and therefore possibly costs you more). In such situations the time taken is greatly increased if there is only one person present to take care of the Auditor. If there are other support staff then at least the Auditor may be able to fill in time by covering other areas.

For the small company in particular this means that good preparation here is essential for a good audit. So, make sure that documentation and records are not too far away physically and that there is a good retrieval system in place. It is usually a good idea to have a good part of what you think it likely that the Auditor will want to see in the room that you have made available. Most CBs will indicate what kind of documentation they will need to see when they confirm the audit.

Pre-Audits

Many companies approaching the Global Standard for the first time like to have a so-called 'pre-audit'. They are unsure that they meet the Requirements and they feel the need to have an external view of their systems before they commit to an audit. This is referred to in the Protocol under Self Assessment of Compliance with the Standard as a 'pre-assessment'. Most CBs offer this service as well. They will usually be conducted along similar lines to a full audit because that is the only way to establish whether the company meets each requirement anyway. However, the company may feel more relaxed about doing it this way, especially as there is not the same pressure to correct nonconformities in time as there will be for the actual audit.

They can be very useful, especially in the early days where a company knows that it is not yet ready. I recall a tomato cannery that asked for a pre-audit and was found to be quite insufficiently prepared for the Standard such that had it been an actual audit they would have been given a Grade D (now 'No Grade'). However, having been able to correct the issues, later that year they had a very successful audit and achieved Grade A.

On the other hand, if you are only contemplating having a pre-audit out of a sense of apprehensiveness you might be better than you think. Try discussing the matter with a CB of your choice and they can advise you. You may not need to spend the time and money on a pre-audit, so think about it carefully. This is especially the case now that we have the Enrolment Programme. At the time of writing, it is not known whether the Enrolment Programme will encourage more companies to go straight to an audit rather than a pre-audit. We shall see.

Note: If an initial audit is going badly it is possible to downgrade it to a pre-audit (*only* on an initial audit) This was sometimes done in the past but this is likely to be less popular and less necessary now that we have the Enrolment Programme. It is <u>not</u> permitted to alter a pre-audit to a full audit once it has commenced.

Note: A CB may not offer advice or consultancy on how to fix or improve things if the same CB is to undertake your actual audit.

Part One

Facilities for the Auditor

Location

Choose a location for the Auditor to be based so that they have a reasonable amount of space and some degree of comfort. There is no need to go over the top with creature comforts but speaking from experience, it is difficult to carry out a good audit when you need to spend all your time balancing a briefcase and clipboard on your knees for several hours.

If you are limited for space and do not have a boardroom, at least ensure that the Auditor has a room where there will be no unwarranted interruptions from passing staff.

Temperature is also important. I was once evaluating a small site where the only available space for a base for the audit was a loft area above production. It was winter and the only heat source was a portable gas fire. Needless to say, I was glad to finish that one and get back into my warm car but, of course, they had limited resources then. I noted that in subsequent years the venue was the same but the heating had improved!

Also consider the location in terms of proximity to records and archives so that if you cannot have much or all of your archive actually present in the room then it is not at the other end of the site. Remember, you are paying for the Auditor's time.

Photocopying

You might need to have photocopying equipment at hand for the audit. At the end of the closing meeting some CBs will leave you with a hand-written summary, which will include any nonconformities that have been raised. The Auditor will retain the original copies, and you will need to take photocopies for yourself. Bear in mind the time that an audit might finish in case certain office areas are locked or at some distance from you by the end. Other CBs might leave you with an electronic version.

Clothing and personal protective equipment

Make sure that you have all necessary clothing and personal protective equipment for visitors and that it is clean and in good condition. Offering an Auditor unclean overalls, hairnets, ear defenders and so forth is not the best way to create a good impression.

Summary

If you follow the ideas set out in this chapter, you should be ready to experience the audit for real. In Chapter 10 you will be taken through the experience of the audit. Before that we will have a brief look at the Directory and any training requirements that you may have.

Part One

Quiz No. 5

(1) What is the typical duration of the audit?

(2) Approximately how many CBs are there worldwide?

(3) How many days does the CB have to complete and upload a report following the audit?

(4) You manufacture canned vegetables and you want to exclude products packed in glass in your audit scope. Is this permitted?

(5) What is wrong with the following audit scope statement?:

The manufacture and sale of dairy products.

8 The Global Standards Website and Directory

Part One

In this chapter, we look at the main website for the BRC Global Standards and in particular the Directory which is used for processing reports and all data concerning the certification process.

The BRC Global Standards Website

The main website for the Global Standards can be found at www.brcglobalstandards.com. In 2011, it was completely redesigned and you should be able to navigate it very easily. For example, if you are looking for information on this Standard, you should click on the menu as indicated:

Click here and you will get a further sub-menu for Food, which in turn has more offerings as follows:

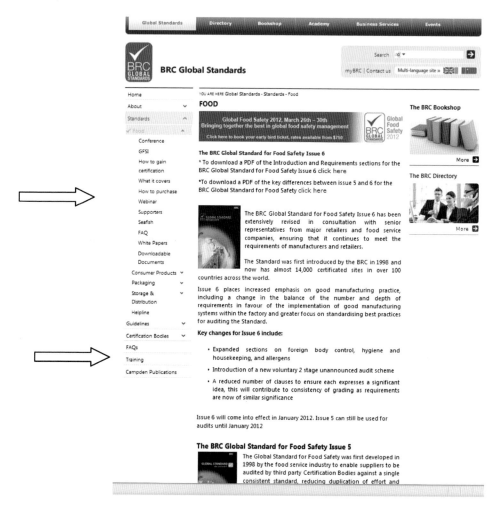

Or, click on to Guidelines like this and see what useful documents you can download, such as the Audit Duration Calculator or the Self Assessment Tool:

It is a really good idea to consult the website regularly because information will be updated there. For example, there are FAQs, and if you click on the Downloadable Documents option, you will be offered the Audit Calculator among several others. If you click on the Training option, also indicated previously, you will find information on all the training courses offered by the BRC. You would do well for the website to become part of your culture and your updating process (ref. Clause 1.1.6).

The Directory

(1) The public area

You can access the Directory in two ways: either by using the second tab from the left at the top of the pages already shown of the main website or directly using the address www.brcdirectory.com. Either way, you will find yourself here:

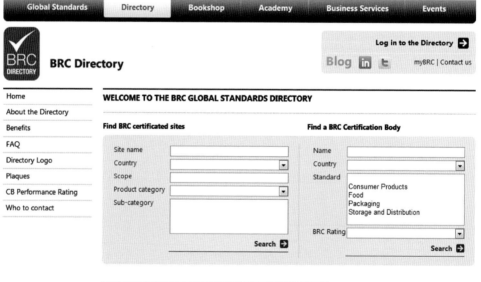

Sitemap | Privacy Statement | Terms and Conditions | Accessibility

Copyright © 2012 BRC Trading Ltd.
Company number 04281617.

BRC Directory is a business of BRC Trading Ltd

BRC Global Standards   BRC Academy
BRC Directory   BRC Business Services
BRC Bookshop   BRC Events

INVESTOR IN PEOPLE

This is the public part of the Directory, where you may search for certificated companies or accredited CBs. It is user friendly, and you will find it easy to follow the guidance given on this page to search for either companies or CBs. If searching for companies, you can search by site name, country, scope, food category or all of these to narrow the search.

For companies, this can be a significant advertisement in that many purchasing bodies will use it to search for suppliers of a particular food category who already have certification.

Once you have found a particular company, if you then click on the name, you will see more detail as to the certification status, including the grade and certificate expiry date. It looks like this:

Site Details

Sea Foods Ltd

BRC site code:	1234567
Contact:	Email: dfinn@seafish.co.uk
	Telephone: 012 555 1234
	Fax: 012 555 2345
Address:	Fish St
City:	Grimsby
Zip code	GY1 1AB
Region/state:	Humberside
Country:	United Kingdom
GPS co-ordinates:	55.07238856756:-1.5555845461238
Telephone:	12 555 1234
Fax:	12 555 2345
Email:	dfinn@seafish.co.uk
Website:	www.seafish1967.co.uk

Certification details:
Standard:	Food Issue 6
	CB Certification Ltd
	Grade B
	4 – Raw fish and fish preparations
	Scope: fresh, chilled and frozen fish portions and bread fish products in flowpack retail packs and 5-kg cases for catering customers
	Exclusion: None
	Issue date: 10/09/11
	Expiry date: 13/09/12

So, even though your customers or potential customers might not have access to your certificate, they can see all the salient details anyway.

(2) The private area

There is also a private part of the Directory, with restricted access to CBs, ATPs and certificated sites. CBs access their part of the Directory for the purpose of uploading auditor profiles and the audit reports. There are also numerous documents available there for CBs and ATPs.

Once a report has been uploaded to the Directory by a CB, an email is automatically sent to you, the company, to invite you to access the new report. You may then give access to any of your customers. Please be assured that the site is very secure and that reports cannot be seen by the public.

The report format is quite sophisticated and contains certain fields of information that are processed by the Directory. For example, an audit that appears to have been carried out by an Auditor who is not registered for the product category will not be accepted by the Directory. Furthermore, the information is used by the BRC to monitor the compliance of the CBs and this contributes to the monitoring systems that we discussed in Chapter 3. As soon as your report is uploaded to the Directory, your entry details are immediately updated for the public page.

Summary

The website and Directory are there for all participants in the scheme: you the company, the CB and the BRC. They are ever changing and they contain much information for you. Make a habit of consulting the website and the Directory.

9 Training for the Standard

Do you feel that you need training in advance of your BRC audit? In this chapter, we look at all the options available to you if you wish for training in the Standard.

A classroom setting can be a valuable experience for this because in addition to receiving professional training, you can meet and network with others from the industry and the cross fertilisation of ideas and experiences is always useful and interesting. You have some options. Many offer training in the Standard these days. You would be best advised to look at either CB's in-house courses or those run and developed by the BRC themselves. That way at least you know that your trainer has real experience with the Standard and has probably been an auditor themselves. In future, there are also likely to be some e-learning options for you.

Some CBs will offer their own in-house training courses for manufacturers. If you are confident in your chosen CB, then this can be a cost-effective option because they can often train several members of staff simultaneously at your premises. The advantage of this route is that your CB may be able to answer specific queries that relate your industry in some depth. The CB must take care that the staff involved in training play no part in any subsequent audit or certification process to preserve impartiality. Make sure you are happy with the course content. The problem here is that this is not a controlled part of the industry, so while it will be cheaper than going for an official BRC course, the content may vary.

Official BRC Courses

Another route is to go for the courses run by and for the BRC themselves. The BRC have developed various levels of training courses for those wishing to learn about the Standard or to become qualified, perhaps to perform audits. At the time of writing, the courses include 1-, 2- and 4-day courses on the Food Safety Standard itself, plus

---

The BRC Global Standard for Food Safety: A Guide to a Successful Audit, Second Edition. Ron Kill
© 2012 John Wiley & Sons, Ltd. Published 2012 by John Wiley & Sons, Ltd.

Table 9.1   Official training courses for the Global Standard for Food Safety from the BRC.

Course title	Duration	Aimed at	Exam?
Global Standard for Food Safety Third Party Auditor	4 days	Those wishing to become Auditors and who do not have a lead assessor qualification, Technical Managers, consultants	Yes
Global Standard for Food Safety 1-Day Conversion Issue 5 to 6	1 day	Technical Managers, manufacturers, consultants	No
Global Standard for Food Safety 2-Day Conversion Course Issue 5 to Issue 6 for Auditors	2 days	Existing registered Auditors	Yes
Global Standard for Food Safety Issue 6 Understanding the Requirements	2 days	New Auditors and new ATPs, consultants, manufacturers. *Note*: To qualify as an Auditor, the person would also need to take 'Global Standard for Food Safety Audit Techniques and Report Writing'	Yes
Global Standard for Food Safety Audit Techniques and Report Writing	1 day	New Auditors	Yes
Global Standard for Food Safety 1-Day Awareness	1 day	Manufacturers, Production Managers	No

courses on the other Global Standards and on related subjects such as HACCP and internal audit. Details on all of them can be found on their website. In future, it is also possible that the BRC will offer courses on other specific subjects. Courses aimed at manufacturers, consultants and those wishing to become qualified auditors at the time of writing are given in Table 9.1.

Note that while each course may be aimed at a specific target audience, delegates may take any course they wish; for example, many Technical Managers take the Third Party Auditor course to gain an in-depth knowledge of how the Standard works. It will be expected that good background knowledge can be demonstrated, for example in HACCP, which is part of the Standard but will not be taught specifically in these courses.

I have been involved myself in the writing and development of these courses for Issue 5 and Issue 6 and I can attest that in developing them the BRC has been at some pains to ensure the quality of the material and the consistency of their delivery by the trainers. They are all quite interactive with many exercises to get delegates involved and improve the training method. They may only be delivered by an ATP and only using the official training materials. So, the advantage of choosing one of the 'official' courses is that you can be sure of this consistency and of getting a full knowledge of the Standard. For some of the courses, you may also have to sit an examination as indicated but that does mean you will get a recognised qualification and a formal certificate. The examinations are all currently multiple choice in style and are 'open book', meaning that you may have your copy of the Standard and all your notes present. There will still be some tricky questions though.

Approved Training Providers

The official BRC courses must be delivered by ATPs. These are individuals who are registered with the BRC, meet their qualification requirements and have undergone BRC training to deliver the courses. A list of ATPs for each country can also be found on the BRC website.

Other Training Requirements

In addition to training in the whole of the Standard, there are key aspects of the Standard that require training to meet the requirements. We consider Clause 7.1 on training in Chapter 17. However, within other clauses of the Standard, there are specific requirements for training that you might also want you or your team to do by using outside help. See Section 'Recurring Themes' in Chapter 4 for a list of those subjects.

There are two specific subjects where I consider training to be essential: HACCP and internal audit. For HACCP, there is much good literature around and many courses are on offer both by the BRC and other organisations. Both HACCP and internal audit are subjects where the individual will benefit from good training courses because they are difficult subjects to teach yourself.

Finding the Right Course for You

As far as CB's own courses are concerned, you need to check their advertising material and websites. As mentioned previously, you should also look at the course content they offer. The ATP-delivered courses are run in two ways. The BRC run them at their venues and you can find the dates and costs involved on the website by either clicking on the 'Academy' tab at the top or directly at www.brctrainingacademy.com. The page looks like this and you should be able to navigate it easily:

However, ATPs are also independent and run their own training sessions using exactly the same material. A register of ATPs is on the BRC website and you should contact the individual trainers to find out their costs and training dates. They may conduct courses on their own premises, so geographically you may find one nearer to you (especially if you are non-UK based). Alternatively, if you have enough delegates within your company, ATPs will usually be happy to carry the training at your site.

Note that the maximum number of delegates for an official course is 12.

Summary

There are many advantages to seeking some external input into your knowledge of the Standard by training and there are some excellent courses available now. They can give an insight into aspects that you might miss and the classroom is a great place for sharing thoughts and ideas.

Part Two
The Audit

In this part of the book, we look at the audit itself starting in Chapter 10 with how to prepare yourself for the structure and nature of the audit; then in Chapters 11–17 we look at every clause in the Standard and prepare you for how to demonstrate your compliance to the Auditor.

The chapters are:

10   How to Survive the Audit

In this chapter, we look at what happens at the audit and how to survive it by knowing what is going to happen and being on your best behavior. We look at the opening meeting, the site audit and the document review. We also look at the vertical audit, which will feature much more now for Issue 6.

Key points in this chapter
Format of the audit
The vertical audit
Health and safety
Listening to the Auditor
Being co-operative

The Format of the Audit

The opening meeting

The Auditor will have made arrangements to arrive at a certain time and will have already agreed to the duration of the audit with you. Let us say that it is to be 2 days. Typically, the Auditor will arrive in the morning on Day 1, say, at 9.00 am and after exchanging pleasantries will go into an opening meeting. This is quite formal as the Auditor has to set out the conditions for the day and let you know the timings.

Very importantly, the Auditor will confirm the scope of the audit and any exclusions from scope. You should be prepared with the correct information for this, remember that the scope must refer to your specific range of products as discussed in Chapter 7. You will have agreed a certain scope at the time of your application but it may be that it needs clarification or more detail.

The opening meeting will include the timing of the site tour and confirmation of the presence of key staff. Do flag up for the Auditor any staff who are on particular shifts or are part time and who may not be available all day (including Human

The BRC Global Standard for Food Safety: A Guide to a Successful Audit, Second Edition. Ron Kill
© 2012 John Wiley & Sons, Ltd. Published 2012 by John Wiley & Sons, Ltd.

Resource/Personnel staff). Generally, Auditors will want to see the factory on the first day but be prepared for the fact that a revisit to a certain part may be called for on subsequent days. Then, the Auditor will also ask about times of shift changes or product changes and will want to be present for those. The Auditor should make arrangements to see any shift changes or product changes where possible. Similarly, they might want to confirm that certain operating staff will be available for interview. They should also flag up that they may want to have a look inside equipment in case engineering staff are needed to be on hand. If this is not your first audit you will be asked to confirm whether there have been any changes to plant and equipment since the last time.

Remember that the most senior Operations Manager (or deputy) must be present at both the opening and closing meetings.

At the end of the opening meeting, you should have time to ask any questions, so then is the time to clarify anything about the sequence of events. If they have not made it clear, a good question to ask is whether the Auditor will let you know about any nonconformities as they arise or whether you will not be informed until the closing meeting. My personal preference is the former because I think it fairer, but some CBs do use the latter technique.

The opening meeting is also your opportunity to give the auditor information on any changes since the application or any future plans for improvements.

The audit

Paragraph 8 of the Protocol sets out the features of the audit that follow the opening meeting:

- Document review
- Traceability challenge (including vertical audit)
- Production inspection
- Review of production inspection
- Final review
- Closing meeting

The auditor has to ensure that 50% of the time is spent in the production facility where they will cover all the GMP issues, seeing production itself and general conditions, interviewing staff, ensuring that they view product or label changes, shift changes and so on. They may also ask for equipment to be opened up.

That, of course, leaves 50% of the time for the rest of the document review (given that they will see some documents in the plant).

Traceability challenge/vertical audit

A significant part of the document review will be the traceability exercise combined with the vertical audit. Auditors are advised to select a product for the traceability exercise, either by purchasing one before they arrive or by selecting a finished product from your site. Note that Auditors are advised to sample product going back no more than approximately 5 months. They will then set you off to complete a traceability test from finished product back through all raw materials. They will also select one raw material for a forwards test, to include a quantity/mass balance check.

In addition, the Auditor will ask for all records associated with that production trace. This can start with raw material specifications and supplier approval, through preparation stages, cooking, packing, storage and dispatch. Along the way it will include process control records, metal detection, weight control and so on and as a consequence can also take in related training, calibration, cleaning records etc. In other words, the vertical audit may become a major part of your document review. Of course, the Auditor is at liberty to go horizontally at any time and will also sample, for example training or cleaning or weight control as separate exercises.

Regarding the maximum 4 hours that traceability is supposed to take (see Clause 3.9.2) the Auditor will have this in mind when carrying out their challenge. However, they will not be expecting you to be able to carry out full traceability and supply all the information required for the vertical audit in that time. If you can do it though – great.

The audit plan

Table 10.1 gives an idealised example of how an announced audit might proceed and the relative timings. Different CBs do things differently. In this example, I have suggested that after the opening meeting, the Auditor may want to have a brief look around to get the feel of the site (and perhaps take a sample for traceability), then look specifically at the HACCP system before going on the main site tour.

The following comments assume a 2-day duration. The colour coding mentioned in Chapter 4 is a guide as to which parts of the Standard the auditor will look at in which half of the audit. Whether this is on the first or second days, or divided between the two is up to the Auditor; however, Table 10.1 is a likely scenario. The traceability exercise for example could be introduced at any time by the Auditor. They might want to test how quickly you can provide the result.

The Auditor is human and will need food and drink and toilet facilities throughout the duration. They should mention lunch breaks in the opening meeting. Do not attempt to drag the Auditor away from the site for a major meal, if the Auditor does not specifically request it, a simple lunch in the office or boardroom is best and will be appreciated. The Auditor will be watching the time constantly and this is not the occasion for lengthy breaks (see also Chapter 7 Section 'Facilities for the Auditor').

Day 1 should finish at about 5 pm. A good Auditor will indicate potential non-conformities as they go. At this point, they might also give you a brief resume of any potential nonconformities found so far, which gives you time to reflect on them for the second day. They might also flag up any documents that they have not yet been able to see to give you a chance to find them on the second day. If the first day has been spent mostly in the factory, the second day will be spent mostly on document review but there is no hard and fast rule.

Note: For an Unannounced audit, the Auditor must start the site tour within 30 minutes of arrival, thus the running order after a brief opening meeting may differ.

Part Two

Table 10.1   Format of a typical BRC audit (timings approximate, refreshment breaks not included).

Timing (min)	Item
	Day 1
30	*Opening meeting*, includes:
	Confirmation of the scope of the audit
	Preferred programme for the days, including:
	▪ production timings including product and shift changes
	▪ availability of key personnel
	▪ arrangements for refreshments
	Confirmation a traceability exercise to be carried out
	Confidentiality/impartiality
	Closing meeting at end
	Questions
30	*Brief site tour overview*
90	*HACCP*
300	*Factory tour*
	Perimeter
	Intake
	Preparation
	Production
	Storage areas
	Staff facilities
	Engineering
10	*Summary of Day 1*
	Day 2
10	*Resume of Day 1*
360	*Food safety and HACCP-based systems, quality management system and audit trails*
	Documents, records, traceability exercise, vertical audit
30	*Return to site for clarification of any issues*
30	Complete Clauses 1–7
30	Time alone to go through nonconformities
30	*Closing meeting*, includes:
	Confirmation of purpose of audit and scope
	Nonconformities summary
	Follow-up details required: corrective actions and root cause
	Report and certificate details
	Information about directory
	Questions

For Day 2, the Auditor should start the day with a resumé of the previous day and briefly outline the rest of proceedings. Much of Day 2 may well involve looking for documents to provide evidence for the various audit trails that the Auditor embarks on. The time this takes will depend entirely on the complexity of your operation. So, the time at which the audit draws to a close cannot be predicted exactly.

The closing meeting

The audit is completed with a closing meeting in which the Auditor will confirm the scope, summarise the day and go through any nonconformities that have arisen. It is a good idea to have all relevant staff present for the closing meeting so that they can hear the summary at first hand. Note that the most senior Operations Manager (or deputy) on site must be present.

The Auditor will explain any nonconformities that have arisen and indicate which clauses they relate to and whether they are critical, major or minor. You will be given a written summary of these and be asked to sign your agreement with them so it is very important that you fully understand them. Take your time and ask whatever questions you need because from the moment the Auditor leaves, the 28-day clock is ticking. It is also a final opportunity to challenge any nonconformities: it may be that even at this late hour you realise that the Auditor has not been given the correct information and that a nonconformity may not be valid.

Do not ask the Auditor for advice on how to correct your nonconformities. Quite simply, they are not allowed to give it and it can be embarrassing for them to have to refuse, (especially as they are usually extremely nice people who would love to help you).

The other thing the Auditor is not supposed to do is give you the result of the audit; in other words, what grade you have achieved. The Auditor should not do this other than to tell you that this decision is made by the CB after technical review. Finally, the Auditor should explain the BRC Directory system and access to reports (see Chapter 8).

So having discussed the format of the audit we will now look at how to create the best environment for your audit.

Keeping the Auditor Safe

Health and safety for visitors is very important. You should make the Auditor aware of any potential hazards on your site and provide them with any personal protective equipment that they might need.

Example

Many years ago, I was in a cannery in Greece that had a smooth tiled floor, which was permanently wet during production times. This was in the days before I took my own safety shoes with me and I slipped on the floor. As it was, I just got soaking wet and a bruise, which meant I preferred standing for the rest of the day! It might have been worse. Since that day I have never entered a factory without my own safety shoes with grippy soles even if the company then wishes to make me change into theirs.

The last thing you want is to cause an injury to the Auditor and next to last is the embarrassment of picking them up off the floor even if they are not hurt. So, keep the

Auditor safe, make sure that they can see where they are going and keep them fully informed of any danger areas.

Preventing Interruptions

We have already discussed the importance of ensuring that the audit flows well and that you are prepared with essential records and documents readily available. It is also important to prevent interruptions to proceedings as it can be most frustrating for the Auditor. Interruptions will only prolong the audit and be an annoyance for the Auditor.

If possible, make sure that the Auditor is based in a part of the factory that is not a general thoroughfare such as an open plan office. Make sure that the receptionist and other relevant staff understand what is going on and they should prevent people wandering in on you or trying to contact you by telephone. Switching off your mobile (cell) phone is a good idea.

Letting the Auditor Lead the Audit

This is very important. It is a real temptation at times to try to lead the audit yourself. Understandably, you may want to show how good you are or how you have improved something. It is OK to mention general improvements at the opening meeting for example, but the Auditor does not want to be dragged off at tangents during either the site tour or the document review just because you want them to see something in particular. The Auditor has much to cover and this will only make things take longer for you. It may even be counter-productive for you in the sense that anything offered up for scrutiny might not be quite as good as you thought.

So, this is a temptation to be resisted. Leave the running of the day entirely to the Auditor. The Auditor will see what they see and you must let them ask questions, listen carefully to what they want and answer them.

Listening to the Question

> I keep six honest serving-men
> (They taught me all I knew);
> Their names are What and Why and When
> And How and Where and Who   (Rudyard Kipling, I keep six honest serving-men)

Auditors are trained to ask questions in a certain way to get the information that they want. Where they want to move things on and get a simple picture of something they will ask closed questions. These are questions where only a 'yes' or 'no' answer can rightfully be given. Interviewers of politicians use these all the time to try to pin down our leaders:

> Prime Minister, are you going to raise taxes or not?

As it happens politicians are experts at not answering closed questions to the extent that even the best interviewers usually give up after the third time of asking. In our situation the Auditor is genuinely trying to elicit a simple response to a simple question:

Do you have cleaning schedules?

For which your answer should be (hopefully) 'yes'. The Auditor may not want to see your cleaning schedules at that time and may have asked the question for future reference. Let the Auditor lead: if they want the evidence they will ask for it. On the other hand, do not make it obvious that you are so wise to the technique that you end up playing a silly word game. In this example, it may be useful and polite to indicate where the cleaning schedules are available, ('Yes, they are kept in the Hygiene Manager's office') thus inviting the Auditor to ask for evidence.

For much of the time the Auditor will want to delve more deeply and draw you out. Then, they will ask open questions which cannot be answered by 'yes' or 'no'. They all start with Kipling's honest serving-men:

Who are the members of your HACCP team?
How do you know that your staff understand your Quality Policy?
When do you change your EFK lamps?
Why have you decided that metal detection is not necessary for this process?
What procedures do you have in place in the case of equipment failure?
Where do staff store their outdoor clothes?

I would say that the 'how' question is the most difficult to answer (although the 'why' question is also a good one) and this is probably how many of the Auditor's questions will begin. These questions draw information out of you, so be prepared for this kind of question with the right information. A good Auditor will develop a line of thought or an audit trail based upon your answers. As always, when the Auditor wants to see evidence they will ask for it, so try to give the Auditor exactly what the question demands and do not supplement with surplus information and evidence. This can often get in the way. In short, let the Auditor lead.

Knowing When to be Quiet

If A is success in life, then A equals x plus y plus z. Work is x; y is play; and z is keeping your mouth shut.   (Albert Einstein, *Observer*, 15 January 1950)

There will be many times during the audit when the Auditor needs a quiet time to review notes or read through documents. These can be lengthy and feel like awkward moments for you. Resist the temptation to break the silence with idle chat; the Auditor does not want too many distractions. It is difficult and occasionally you might excuse yourself for a few minutes to give both of you a little space.

Part Two

Being Positive and Co-operative

The Auditor will be well aware that they are not the most welcome visitor of the year but you should not make them feel unwelcome. They are experienced in industry and most are quite sympathetic to the situation that you are in, indeed many have experienced it themselves.

If you try to place obstacles in the way or make the Auditor's life difficult you will be doing yourself no favours. Be positive about the audit. Consider that the Auditor is in fact performing a service for you from which you will benefit. You will not only come out of it with a certificate (hopefully), you will be assured that your systems are in good order for your own due diligence and perhaps improve them through corrective actions.

Your audit will be a better experience without antagonism of any kind, so in a nutshell: be nice. Even at its worst, it is a learning experience.

Consultants

There are many consultants who offer a service to prepare you for a BRC audit. Most do a fine job and have much experience in doing so. Indeed, I know many. From the Auditor's point of view, a certain amount of trust can be built up with good consultants. However, there are a few who do not have the right experience for the job and worse, make life difficult for the company and the Auditor by arguing points that are clearly issues that they could have addressed. When such a consultant is faced with the reality that they have not done a good job for the company, they may react too defensively during an audit to the detriment of the process. I have been present on such occasions (see Example to Clause 2.1.1).

The use of consultants is understandable, especially for small companies who do not have all the necessary expertise. However, it is wrong to engage one and let them work in isolation because you feel that you have nothing to contribute. You should work with your consultant. Make it clear that they are not solely responsible for the outcome but that you consider it a team effort. That way they do not need to be overly defensive when nonconformities are raised, you will get the best out of them and situations like the aforementioned are avoided.

At the time of writing, there is some talk at the BRC of having a register of qualified Consultants who have undergone the official training courses.

Remembering the Obvious

We can all make mistakes under pressure, so take a breath and have a mental checklist at various points in the day so that you do not forget certain things that might give the wrong impression. Make sure that the Auditor has signed in as a visitor and completed any visitor's questionnaire that you have. Remember to keep the Auditor refreshed and offer drinks at regular intervals. While you are asking the Auditor to remove their watch and jewellery, do remember to remove your own. Similarly, remember all the usual protocols such as putting on overalls in the correct order, hand-washing on entry and so on.

Make sure that the receptionist is fully aware of your visitor beforehand and the purpose of their visit. It gives a good impression of a well-organised company if the Auditor can see that they are expected and that the receptionist knows exactly who to contact as you arrive.

Correcting Nonconformities During the Audit

A good Auditor will make you aware during the audit of any issues that are likely to be nonconformities. Thus, you have a chance to think about them and there are no nasty surprises at the closing meeting. If the audit is going into a second day, they may even have a short summary meeting at the end of the first day. This gives you the opportunity at least to make sure that the Auditor has seen all the evidence pertaining to the issue.

However, this is not the same as taking new action to correct a nonconformity during the audit. You may feel that there is an advantage to be gained by correcting nonconformities before the closing meeting. Indeed there is, if the Auditor can verify corrective action during the audit it does mean that there is one less for you to deal with during the 28-day deadline. However, note that the Auditor will still put the nonconformity down on the summary and *it will still count towards your grade*.

May We Offer the Auditor a Gift?

In short, no. It is better not to embarrass the Auditor in this way because they really should not be accepting anything from you for fear of being seen to be compromised. Auditors are forbidden to accept gifts or payment in kind that may cast doubt on the integrity of the audit or certification process.

Summary

Remember to create a good atmosphere for the audit by being polite and co-operative. Look after the Auditor's safety and comfort. Avoid interruptions and have all the right staff available. Listen to the questions and let the Auditor set the pace and lead the audit. Your professionalism will be obvious and give the Auditor much reassurance.

So, let us now have a look at the Requirements.

Part Two

11 Clause 1: Senior Management Commitment

The Requirements begin with an emphasis on management commitment and, in particular, senior management involvement in all aspects of food safety. This theme is continued through the Standard but here the foundations of this theme are laid. You will see that while one of the key words here is commitment, another is communication. I have decided to take a sideways look at Clause 1 by changing the running order, mostly to suit myself but also to keep you awake! Note that Section 1 is mostly colour coded green and is therefore expected to be mostly audited during the document review part of the audit.

Clause 1.1: Senior Management Commitment and Continual Improvement (Fundamental)

Key points in Clause 1.1
Policy
Commitment, resources and improvement
Objectives
Management review
Timescales and planning
Root cause
Communication

The Standard starts with a SOI, which is a *Fundamental Requirement*.

The thinking behind this clause is to emphasise that Quality and Food Safety systems are not purely the domain of the technical staff. This may have been the idea in the past but such thinking is not acceptable in modern food production.

What is needed here is involvement in, and full commitment to, the Global Standard by the senior management of the company. One of the key words here is that they need to demonstrate this. The Auditor will be looking for evidence of this throughout the audit; possible examples are senior management involvement in Management Reviews, HACCP, Internal audits and so on. One of the key aspects is resources

The BRC Global Standard for Food Safety: A Guide to a Successful Audit, Second Edition. Ron Kill
© 2012 John Wiley & Sons, Ltd. Published 2012 by John Wiley & Sons, Ltd.

which is underlined in Clause 1.1.5 You must show that the senior management have adequate resources which are applied to the food safety systems.

Furthermore, an important aspect is that of improvement. The Auditor will be looking for evidence that your senior management is committed to improvement and that there is a process here where opportunities are identified and any improvements are documented.

Clause 1.1.5

This Requirement is purely about your resources and requires that you provide sufficient resource, both financial and human to produce safe food, meet the Standard and implement a HACCP food safety plan. It used to be the first clause of the Standard and I think it sufficiently important to start with even though it appears later in the section now. Conversely, from the Auditor's viewpoint, it is not usually possible to judge this until near the end of the audit. Usually, if very few problems have been found with any of the systems, then it will be inferred that the resources provided are sufficient.

On the other hand, if there are problems that are clearly caused by insufficient staffing or funding for example then this clause may lead to a nonconformity.

Examples

A medium-sized meat products company had a FSQMS that was mainly run by the managing director. As there was no real backup for him in the event of absence, the Auditor considered that they were under-resourced in this area.

A jam factory had problems. There was inadequate Quality or Technical personnel resources available on site; partly due to the recent resignation of previous Quality Manager in August, not yet replaced.

Thus, a nonconformity was given in each case.

In a company brewing and canning beers, the Auditor considered that the company resource to maintain and improve the FSQMS & HACCP systems had not maintained parity with the on-going site upgrade project and the large increase in volume output that had resulted recently.

In this case, the Auditor gave a major nonconformity as the situation was acute.

Clause 1.1.7

Those resources must stretch to purchasing at least one copy of the Standard. This could not be more straightforward. If you do not have an original copy of the current issue of the Standard available to show the Auditor, a nonconformity will be raised, and I am afraid it is likely to be a major.

You know what you have to do.

Clause 1.1.2

Here the Requirement is concerned with senior management's involvement with the food safety and quality objectives. This may be considered to be in three parts. Firstly, quality objectives have to be established and documented. They must also be communicated to the relevant staff. Finally, you must monitor objectives and report results to senior management on at least a quarterly basis. Note that if you are part of a group of companies this must be reported to site management not central management.

So, what sort of objectives should be set in the first place? The Auditor will be looking for something more precise that the general objectives set out in your Quality Policy. For example, it may include the following:

- Reducing consumer complaints
- Fewer nonconforming product reports
- Fewer nonconformities raised at audit
- Improved customer satisfaction
- Improved pest control
- Improved weight or volume control

There are many others you could think of but they can all be quantified and they must be documented. Importantly, the Auditor will be looking for objectives that have been properly thought through. They must be measurable, achievable and sensible deadlines should be set.

Part Two

Examples

A relatively small bakery had a fully documented FSQMS. They were able to demonstrate senior management commitment to the cause. However, they had no quality objectives, targets or KPIs established at all.

At a chocolate confectioners, the quality and safety objective for the current year were not clearly defined.

A nonconformity was raised in both cases.

How do we communicate objectives to staff? There are several ways companies can do this. It can be through training, memos or newsletters, information noticeboards and so on. These objectives must be monitored and reported. In other words, you have to demonstrate that they are a part of your continuous measure of your performance. Note that you must do this at least every 3 months.

> ### Example
>
> A speciality bakery had set quality objectives and these were documented, but there was no reporting of performance against them.
>   Thus, a nonconformity was given.

It is also important to note that the objectives set must form part of your management review (see 'Clause 1.1.3').

Clause 1.1.3

Management Review is a cornerstone of your system. This is important because it demonstrates that the FSQMS and HACCP systems are not just the domain of the technical department but are owned by all, including the senior management. Those familiar with Issue 5 of the Standard will know that Management Review took up five separate clauses in this section and this clause is an example of the merging of clauses which happened in the review to Issue 6. The Auditor will be interested in the attendees of the meeting and you should ensure that the senior management of the site attend. If you have a separate head office, it might also be necessary to include some head office staff as well.

Commonly, Management Review is an annual event which is the minimum requirement. They could occur more frequently, but they are not to be confused with a simple management meeting that might occur weekly. This has been a common finding of Auditors in the past.

Certainly, the review will manifest itself as a meeting and this should have a set agenda (see the following text).

> ### Example
>
> At a produce packer, there was no management review record available and furthermore scheduled dates for review had been missed.
>   A nonconformity was given.

This review must look at performance against the Standard and against any objectives set to meet Clause 1.1.2.

Example

At a cheese manufacturer apart from complaints monitoring, quality objectives were not documented in management review.
    A nonconformity was given.

This is the first of many clauses that give a list of items that you need to include in satisfying the Requirement. Thus, in your review process, you need to include *every one* of the following (see clause for exact wording):

- Review of previous management review meeting minutes, action plans, time frames.
- Internal, second-party and third-party audits.
- Customer complaints, results of any performance reviews by customers.
- Incidents, out-of-specification issues, nonconformance issues, status of preventative and corrective actions.
- Reviews of management of HACCP.
- Resources.

Of course, you are free to discuss any other issues such as the effect of legislation changes or you might use the review to flag areas for improvement.

Missing out any of the specified items will lead to nonconformities being awarded. In a sense this requirement is a 'gift', so plainly it is set out, so it is all the more surprising that nonconformities are commonly awarded against this clause. Here is a selection of nonconformities, each from a different site:

Examples

Internal audits not discussed at management review meetings.

The objectives of the previous review not reviewed.

Management reviews did not include in the agenda a review of previous minutes or recommendations for improvement.

Management review did not include all the requirements of the clause.

Management review did not include previous minutes or customer performance reviews.

An agenda is a wonderful thing. Once written, it should dictate the content of any meeting. The simple way to meet this clause is to set up an annual agenda that at least includes all six of the aforementioned subjects and use this as a basis for all your management review meetings in future. The Auditor will be looking for all of them in your management review records. Records are key here because the records of

Part Two

this meeting are expected to be used to review objectives. If you have a major review annually, then it should not be a problem to keep a full record of this. However, some sites set out to have more frequent reviews, say, monthly. If you set a schedule, you must meet it. If your records let you down, you could receive a nonconformity.

> **Example**
>
> A manufacturer of sugar confectionery had a procedure for monthly management reviews. However, they had failed to record any reviews for the 3 months from May to July of that year.
>   Therefore, a nonconformity was given.

It would be expected that regular meetings would be minuted and the minutes distributed to those present at least. Do ensure also that any other appropriate staff are circulated with the results of review. The Auditor will expect to see such minutes documented and circulated in this way, verbal communication may be acceptable at a certain level but remember that the Auditor needs to see something to satisfy the requirement.

> **Example**
>
> A manufacturer of hamburger meats and sausages had a management review but the review and actions agreed were verbal and not documented for action by designated staff. The Auditor was not satisfied that there was clear evidence of the information being made available to the appropriate staff.
>   A nonconformity was given.

Similarly, it should be clear that any actions are allocated to particular members of staff.

> **Example**
>
> A processor of prawns and other seafood had minuted their management reviews but had not assigned actions to any particular members of staff. The Auditor therefore considered that the results of the review had not been clearly communicated.
>   Thus, a nonconformity was given.

Finally, all actions must be implemented within the agreed timescales and the records updated accordingly. Remember that this includes not only corrective action but may also require improvement action. The Auditor will expect to see documentation completed and actions recorded as completed.

Clause 1.1.4

This clause concerns addressing food safety, legality or quality issues. An 'issue' may be anything from an incident such as a product recall situation to perhaps a series of nonconforming products that requires communicating through your company. The idea is that senior management shall be informed of any such issues and that a meeting shall take place at least monthly to effect this. Clearly, you will have to provide evidence of such meetings for the Auditor and they will be looking for records of them and of any action taken on issues.

Clause 1.1.6

You must demonstrate to the Auditor a positive method of keeping up to date with legislation, issues, technical developments and Codes of Practice. You need a system in place for this and not a passive reliance on picking up information for example from your major customers (several of the UK retailers are very good at reminding you of issues, of course).

One good way is membership of trade or technical organisations within your country. Very often, these will issue newsletters on important developments within your product range. Be aware though that you need to keep abreast of any general legislation, food related or otherwise, which affects all of the food industry.

Some small companies rely on internet searching for their information, for example in the United Kingdom such companies may gather much information from the FSA website (www.food.gov.uk). However, it is important to be able to show the Auditor a positive method; so perhaps if you use this method, you will need to keep records of your logging on to the FSA site.

Note also that the Requirement is for you to be aware of this information in any other country where (a) your raw materials originate or (b) your products are sold. So, if you are importing raw material or exporting finished product, you will need to show evidence of your method of updating the relevant legislation in your supplier's or customer's country.

Example

At a mushroom canner, the company obtained no information and updates on legislation, guidelines, codes of practice and so on. During the audit, we observed that the legislation on packaging and pesticide residue for country of destination was not accessible.
A nonconformity was given.

The Auditor will be looking for a pro-active system that is formalised.

Clause 1.1.8

This clause applies to those already certificated. If the CB is aware that the current audit is late for no justifiable reason then the Protocol dictates that a major nonconformity will be given, against this clause. Assuming all is well on this occasion, the Auditor may want to see evidence of planning ahead 12 months (or 6 if appropriate) to ensure that the subsequent audit is carried out on time. For example, do you ensure key staff holidays avoid audit dates?

Remember: If you delay your audit without justification, you will get a major nonconformity. *That is halfway to a Grade C!*

Clause 1.1.9

This Requirement concerns the staff present on the day of the audit. It is considered important that senior management take a full part in the audit process; hence, your most senior Production or Operations Manager on site must attend the opening and closing meetings.

The Auditor will establish who those persons are in the opening meeting and you should ensure that they are present. In addition, you must also ensure that Department Managers or deputies are present for the audit.

This should become another part of your preparation and planning.

Clause 1.1.10

This requirement applies to *all* nonconformities raised at your previous audit, you must ensure that have been effectively corrected and that the root cause has been addressed.

Example

At a confectionery factory, nonconformities from the previous audit were not verified as actioned due to significant personnel changes.
  A nonconformity was given.

The Auditor will be aware of corrective actions submitted previously but will need to confirm those with you during the course of the audit that the root cause action plans you submitted were effected.

Note: If you change CBs, you must provide your current CB with the previous report.

Clause 1.1.1

Here, we have what should be a relatively straightforward requirement. The clause requires you to have a Policy Statement stating your commitment to produce safe and legal products to specified quality. It must be clearly documented, signed and communicated to all staff. These documents are still widely referred to as Quality Policy Statements even though quality is only one aspect.

By 'signed', the Auditor will be looking for the document to bear the name and signature of a senior representative such as the Managing Director.

Example

A wine supplier was found to have no signed or otherwise authorised copy of a Quality Policy Statement.
   A nonconformity was given.

This clause is generally well understood by companies and it is relatively rare for a company not to have a 'Quality Policy Statement'. However, you must ensure that it also has a food safety content. Most nonconformities arise out of shortcomings in the content or in its application as shown in the following text. The Auditor will be looking at the wording of your policy to see if it includes the right references.

Example

A manufacturer of canned apples had an authorised Quality Policy Statement, signed and dated. However, it made no reference to any intention to produce 'safe and legal' products.
   A nonconformity was given.

The Requirement is also concerned with how you communicate the contents of the policy to all staff involved with product safety, legality and quality.

Examples

A processor of frozen sweetcorn had a Quality Policy Statement which met all the requirements. Including seasonal workers they had approximately 480 staff. The policy was authorised, it stated their intentions to produce safe and legal product, and it was communicated to management and understood by key staff. However, there was no system in place to ensure that it was communicated throughout the company to all staff involved in product safety legality and quality.

Similarly, at a confectionery factory, there was no evidence that Quality Policy is communicated to all staff.

A nonconformity was given in both cases.

Part Two

What the Auditor will be looking for are examples of induction programme, further training, review meetings and communication generally to ensure that this level of staff are provided with the right kind of information and understanding of these matters. It may be necessary to have staff sign off their understanding of the policy as a formality.

Many also display their Policy Statement at various locations on site, for example reception area, staff rest room. You should also ensure that temporary staff and those working off site, for example delivery personnel are made aware of it.

Example

At a manufacturer of bottled mineral waters the Auditor found that while there was a Food Safety and Quality Policy Document, and its existence was referred to in the staff induction, the contents were not conveyed to all staff involved in product safety legality and quality.
  A nonconformity was given.

Once a document is on a noticeboard, it is easy to forget about it though.

Example

At a confectionery factory, the Quality Policy on display in the reception area, was an obsolete version (Current Issue 17–2-11, display version 17–9-08).
  A nonconformity was given.

The Auditor may well speak to staff during the factory tour and ask them if they are aware of the Food Safety and Quality Policy. In the European Union, there are a growing number of food companies that are multi-lingual because of the increased movement of workers across national borders. A potential question for an Auditor to ask in this situation is whether all the language groups within a site can demonstrate understanding of the policy.

Clause 1.2: Organisational Structure, Responsibilities and Management Authority

This SOI is an objective to have an organisational structure whereby responsibilities, job functions and reporting relationships are clear and documented. Note that the clause, like several in the Standard uses the phrase 'product safety, legality and quality' in defining the activities to be managed.

Key points in Clause 1.2
Organograms and responsibilities
Deputies
Work instructions
Dealing with languages

Clause 1.2.1

This clause is one that has resulted from a merging of three old clauses, so I will deal with it in three parts. Regarding the organisation, there are one or two ways of presenting this information but the most popular among companies is the organisation structure diagram or Organogram. This is a requirement of this clause. It is a simple document to issue and is favoured by Auditors for its simplicity and accessibility. A good diagram would include all the key positions with lines indicating the hierarchy and reporting relationships. Elsewhere, you can have detailed information about what each position entails (see Clause 1.2.2).

Note: It is usually better to name only the titles of positions and not put in actual names of individuals. If you put in names of persons you will only have to update the document each time someone leaves or changes position.

In the second part of the clause the word 'responsibilities' is the key one. You are required to ensure that managers, that is those with responsibility for product safety, legality and Quality Systems are clearly allocated their responsibilities. It means not only their obligations to you the employer in carrying out their duties, but also their legal responsibilities as food handlers. There could also be a link here with the requirements of Clause 7 and staff training.

The final part of the clause is one that companies are sometimes tripped up by. You are required to have arrangements in place for the absence of key staff. In simple terms, you need to ensure that each key member of staff has at least one deputy who can take up such duties when necessary. The Auditor will want to see formal arrangements for this for designated staff to cover absences. What is unacceptable is where the arrangement is simply that 'management will allocate a deputy in the event of absence'. There needs to be a mechanism in place before the event so that absences are covered as seamlessly as possible.

Example

A tomato cannery had an organogram and reasonable staff training systems and job descriptions, but had not set any arrangements in place, such as named deputies, for the absence of key staff.

Thus, a nonconformity was given.

Note: For some specific subjects such as management of pest control, you are required to have defined responsibilities (see Clause 4.13.3) so this could be indicated here.

Part Two

Clause 1.2.2

This clause is about awareness of your responsibilities and your working procedures and work instructions. Here we are looking for detailed procedures on all aspects of production that affect food safety, legality and quality (that phrase again). The Auditor will expect to see documents that have been well formatted, are clear, accurate and controlled (see Documentation Control later).

Example

A poultry processor had a documented procedure for their internal audit system. The procedure described the frequency of audit as 6-monthly. This was inaccurate as in reality the systems were audited annually. In this case, it was the procedure that was incorrect not the activity.

Therefore, a nonconformity was given.

Uniformity of format is useful, not only for the Auditor but probably for personnel as well. A 'ragbag' of work instructions written in an uncoordinated way over a lengthy period may not impress, although the key here is that they are clear and comprehensive. Each procedure should state exactly what the task is, who is responsible for carrying it out and a sequential list of instructions. It should then make clear what to do if an irregularity occurs which might lead to nonconforming product. All your essential procedures should be documented in this way.

Examples

At a produce packer, there was no work instruction covering CCPs monitoring or the checking of product at reception for temperature.

A small manufacturer of frozen dumplings neglected to have written work instructions for either their cooking or freezing processes. As these both happened to be CCPs this was a relatively serious matter.

A nonconformity was given in each case.

The other important point in this clause is the communication of the work instructions to the relevant staff. Part of this may come up under training, but the Auditor will need to be assured that the people carrying out each task have access to the document and they may pick on individuals during the factory tour to ask about their awareness of the procedures. They will also want to see tasks in action.

There is also the issue of language. Because of the increased mobility of workers, in any factory there may be several languages spoken. It is necessary to make sure that each language is appropriately represented in work instructions.

Summary

The essence of this set of clauses is to create a culture whereby your commitment to food safety and quality is created at the very top and is felt all the way through all levels of management and operators. Ensure that you are properly resourced and can demonstrate your commitment and have good communication channels. Make sure your audit happens on time, make sure all your root cause action plans have been carried out and buy an original copy of the Standard!

This section is also about being structured in your organisation so that everyone knows what they are about. It is all about communication and documentation. Make sure that you have deputies in place for all functions and that everyone understands their job function and work instructions.

Quiz No. 6

(1) What is the result of delaying your audit?
(2) How often must you meet to discuss issues?
(3) Who should sign your Policy Statement?
(4) Objectives shall be documented, communicated and what else?
(5) What is the first clause of the Standard in which the term root cause appears?

Part Two

12 Clause 2: Food Safety Plan – HACCP

HACCP has become the essence of food safety control, certainly in most countries. The logic of HACCP is irrefutable and it is no surprise that it is one of the cornerstones of the Global Standard. Since you are reading this book with a view to being certificated, I will assume that you have some knowledge of HACCP principles. If not, you should consider attending a training course on the subject, of which there are many on offer, as discussed in Chapter 9.

Note that this section is predominantly green with a little peach; however, no doubt the Auditor will look at many aspects during the site tour as well.

Clause 2.0: Food Safety Plan – HACCP (Fundamental)

> **Key points in Clause 2.0**
> Codex Alimentarius
> Retaining documents

The first SOI within this clause is also a *Fundamental Requirement*. It is saying primarily that you must have a HACCP system that forms the basis of your food safety control systems.

There are two further points to be conscious of in this SOI. Firstly, there is a reference to Codex Alimentarius HACCP Principles (Basic Texts: Hazard Analysis and Critical Control Point (HACCP) System and Guidelines for its Application). You must demonstrate that the HACCP system has been prepared by following the Codex principles.

So, what will the Auditor be looking for here?

Well, clearly, you will need a documented HACCP system available for examination. This usually means a HACCP Plan with all the requirements of Clauses 2.8–2.11 documented. However, the Auditor will also want to see any records that you have retained that show how you put the system together so that it can be confirmed that

Codex principles have been followed. Such records will also let the Auditor follow your trains of thought generally. For example, you should retain original documents that show all the potential hazards that were considered, whether they were considered significant and were subsequently used to determine CCPs.

Note: It always looks slightly unlikely to the Auditor when they see that every hazard considered at the outset made it through to the final list of CCPs. It is not impossible, of course – but watch them start making notes at this point!

Indeed it can be frustrating for an Auditor to be presented with only a final HACCP plan, because no matter how comprehensive the plan looks, just like the maths teacher at school (in my day, at least), a good Auditor wants to see the 'workings out'. If you have retained documents such as your original hazard analysis, lists of potential hazards (even those you have discarded), lists of possible global hazards and prerequisites, original notes and decisions on CCPs, notes of early meetings: all of these must be available for the audit. Even the scruffiest notes can go down well with the Auditor here. It gives the Auditor a chance to assess the depth of the original study.

Little better than no HACCP is the perfectly presented HACCP plan that seems to have appeared out of nowhere, especially if it has been put in by someone from outside. The words 'off the peg' come to mind for the Auditor and alarm bells may start to ring. A HACCP system should be the core of your technical approach to the safety of your products and if the Auditor feels that it is purely an 'add-on' you may be given nonconformities.

The Auditor is looking for a HACCP system that truly reflects your own production processes and potential hazards and not a package that you have bought in and that sits on a shelf. Here is an extreme example of what can be wrong.

Example

A site producing bottled mineral water had a HACCP system on paper which included references to the Campden & Chorleywood Food and Drink HACCP Practical Guide, as was, and BSDA guidelines to good hygiene practice for bottled water.

However, there was no evidence that the Codex Alimentarius principles had been followed. There was some reference to legislation but the hygiene regulations quoted were out of date.

The HACCP team had signed off a review but there was no comment on the lack of proper risk assessment. There were no critical limits stated for CCPs. There was no monitoring of CCPs, no corrective action set and so on.

Most significantly, there was no evidence of the site using HACCP as a tool at all.

They failed against several clauses under HACCP including the Fundamental Statement of Intent and resulted in being un-certificated.

Similar problems can occur when a company has recently become part of a large group and the HACCP is handed on to them by a central function. Sometimes, when this happens, parts of the system are not fully implemented.

> ### Example
>
> A company that was part of a large dairy group had been 'given' a group-wide, generic HACCP study but this was not yet implemented on site.
>
> The company received several nonconformities under specific clauses in Section 2 and a nonconformity under this Clause 2.0 because the new system as a whole was not fully implemented.

So, let us look at how it should be done.

Codex Alimentarius

You can download a copy of the Codex Alimentarius Basic Texts on Food Hygiene (2003 – and no doubt any subsequent editions) from www.codexalimentarius.net and search for 'HACCP'.

Note: And the good news is: it is free of charge.

Any experienced, trained Technical Manager is likely to be quite aware of the content as it will form the basis of most HACCP training courses, but it does no harm to go back to the source material occasionally. Codex suggests seven Principles for HACCP and in addition adds five prior stages in preparing a HACCP system giving 12 steps in all. As mentioned previously, the Global Standard now follows these steps in logical order.

Part Two

Other Aspects

Aside from Codex, you need to have reference to relevant legislation, codes of practice or guidelines specific to your products. Clearly, this will include all the general food safety legislation that applies in your country, but also specific requirements, codes of practice and guidelines for your products. This is covered specifically in Clause 2.3.2. Remember that the Auditor will know something about your industry so you will need to ensure that you have kept up to date with any such material and refer to it in your HACCP.

Clause 2.1

The single clause under Clause 2.1, that is 2.1.1 concerns Codex Step 1: Assemble a HACCP Team. (The clause derives from three clauses in Issue 5).

Key points in Clause 2.1
HACCP team
The team leader: expertise/competence/knowledge
External expertise
Day-to-day management of HACCP

Clause 2.1.1

HACCP really is a team game. If the resources are available, the best situation for the Auditor to see is a broad spectrum of team members: the so-called 'multidisciplinary team'. For example, this might include the Technical Manager, Production Manager, engineer, Hygiene Manager and others. The Auditor may expect to see them listed in the early pages of the HACCP system but beyond a list of names or positions, the Auditor will want to see evidence of their actual involvement. For instance, you could have examples of minutes of initial meetings, taking part in HACCP audits, minutes of review meetings and so on: in short evidence of constructive involvement by all. The Auditor may want to question one or two team members on certain aspects so it is a good idea to make sure your HACCP team is on site and available on the day of the Audit.

Codex does say that it may be possible for an individual to implement HACCP in house, especially where resources are not so freely available. But what Codex also says in this case is that the individual should be supported by expert advice from outsiders such as the local authorities, consultants, trade bodies and the like.

The subject of HACCP training is not specified in this clause, but rather that those involved in general can demonstrate competence. The team has to have specific knowledge of HACCP but how this is come by is left open. In reality though the Auditor will usually feel much more comfortable in judging this situation if there is evidence of training. Specifically regarding the Team Leader and their competence in HACCP principles, it is one thing to have good qualifications but there is more to it than that. The Team Leader has to demonstrate competence and experience in understanding HACCP. The Auditor may have seen all your Team Leader's impressive qualifications, which hopefully you had available. They may look good but if there are serious flaws in the system they may take the view that competence has not been demonstrated. Competence is truly demonstrated by presentation of a thorough analysis and a competent summary. If training has been given to team members in house, then you will need to have evidence that the team members have successfully completed the training. It may not be necessary for certain team members to have quite the depth of knowledge that the Team Leader has, so in-house training may well suffice. However, if the Auditor feels that sufficient evidence of competence is not demonstrated or that there are deficiencies in the HACCP system, which would be answered by better training all round, then this situation might lead to a nonconformity.

The best way to demonstrate compliance here is to have a good system and meet all the other requirements of Section 2. The Auditor may be concerned if the Team Leader is self-trained and carries out the in-house training for the other team members. The lack of input from the outside world is not likely to inspire confidence. HACCP can be a difficult concept to grasp and very unfocused systems may result from this kind of scenario. Remember also that your team members may have changed recently and that an up-to-date list of competent members is essential.

Example

A poultry manufacturer with a well-established HACCP system that had been reviewed several times was evaluated and generally all was well. They had a HACCP team comprising five members of staff.

However, since their previous Audit they have had an unexpectedly high turnover of staff and have three new members of the team. No provision had yet been made to carry out HACCP training of these three and they had no previous experience of HACCP.

Hence, a nonconformity was given against this clause.

Regarding the rest of the team, the Auditor will judge much from the content of the system itself, about their knowledge of HACCP. However, consider the following example.

Example

A grader and packer of potatoes was a very simple operation with few demands of a technical nature. Therefore, they did not have much technical expertise in house. While they had a HACCP system drawn up, the Auditor considered that they did not have the appropriate in-house experience to develop and review the HACCP system.

Therefore, a nonconformity was given.

In the case of very small companies the team may also be small, for example two or three. Provided it satisfies the multidisciplinary requirements, this can be acceptable.

The final part of the clause concerns the use of external expertise. If you have a consultant working for you, and sometimes they become the Team Leader, it is essential though that your staff can be seen to be managing the system on a day-to-day basis. The Auditor will be uncomfortable with the situation if it appears that the consultant is running the HACCP but they are only on site once every 6 months. A HACCP system needs to have life breathed into it on a daily basis.

We discuss consultants in Chapter 10 but if you do use one it can be a good idea to have them present for the Audit as well. For example, the Auditor may want to see evidence about their qualifications, something that companies often fail to do properly.

Basically, the Auditor will need to be convinced that either there is sufficient expertise in house, or it has been properly sought from outside. If you are a Technical Manager or Team Leader, be ready to show evidence of your own qualifications and experience (certificates and so on) and that of your team. If you are using a consultant, ask them to bring theirs.

This can be a tricky one for the Auditor.

Part Two

> **Example**
>
> I once had to chair a difficult closing meeting involving a very opinionated consultant who had put together a HACCP system for a company. He had clearly been in industry for many years and considered that he did not need any formal training in HACCP as he had been applying it for so long. In fact, he was outspoken on this subject, to the extent that he poured scorn on the necessity for training someone of his years in a system, which in his opinion was pure common sense.
>
> However, they had two major and three minor nonconformities in the HACCP section. One of the majors was that they had not identified any CCPs. Given the nature of their process, there should have been some.
>
> As the HACCP system had several significant flaws, and the lack of any CCPs was a major howler, I had to tell him and the company that I did not agree that the consultant was sufficiently competent and gave a further nonconformance under this clause. His only possible corrective action was to go back to 'school' and learn to apply HACCP principles properly.
>
> As I say, a difficult closing meeting.

Clause 2.2

This is a new addition for Issue 6, an entire clause concerning the subject of prerequisites within the HACCP section of the Standard. It is the only significant change to the Section.

Key points in Clause 2.2
Prerequisites in relation to HACCP
Review

Clause 2.2.1

So, what are the prerequisites?

> The production of safe food products requires that the HACCP system be built upon a solid foundation of prerequisite programs.   (US National Advisory Committee on Microbiological Criteria for Foods, *Journal of Food Protection*, 1998)

Essentially, it refers to all those topics that could come under the heading of GMP and that are really global issues in any food manufacturing site. You could be forgiven to wondering why this clause is there at all. After all, every one of the bullet points listed here is surely covered in the relevant parts of the Standard. The bullet points here, please note, are intended as guidance.

We think of prerequisites as areas requiring control but which do not relate to a particular process step. In other words, they are global as regards your site. Pest control is a good example. Therefore, they are not part of the HACCP process where

CCPs are identified. The idea here though is that prerequisites are identified as part of the overall FSQMS and that they are clearly documented and included within the development of HACCP and in HACCP reviews.

The Auditor will be looking to see if your prerequisites, which may include this list and more, have been taken into account during the development of your HACCP system (another reason to keep all your early records). As far as reviews are concerned, the Auditor will also want to see reference to prerequisites. Remember also that some of the items listed in this clause such as allergen controls, could indeed relate to a process, depending on the nature of your production, and could therefore be part of a HACCP plan itself. Finally, you must clearly document control measures for prerequisite programmes. It is likely that the Auditor will cover this subject when looking at those items specifically. See also Clauses 2.12, 2.13 and 2.14, which also concern prerequisite programmes including the subject of review.

Clause 2.3

All the clauses under Clause 2.3 concern Codex Step 2: Identify the Product.

This should be an easy one. Remember though that in identifying the product the process is fundamental to it and that the process itself may be the defining character (Clause 2.3.1).

Key points in Clause 2.3
Scope of the plan
What to include
Food safety characteristics
Hazards
Legislation/codes of practice
Keeping it updated

Clause 2.3.1

To meet this clause you must define the products and/or processes that are included in your HACCP system. Clearly, from the Auditors point of view, this must include all the products and processes that you have decided to include in your scope (see Chapter 7).

This should be a simple matter in the early stages of your HACCP development and the Auditor will need to see the evidence of the products and processes defined in the final documents.

The importance of this in developing your HACCP system should be obvious and only by considering all such potential sources of hazards could your HACCP system be (and be seen to be) comprehensive.

Therefore, the Auditor will be looking for well-developed documentation in the preamble to the final HACCP plan where it can be seen that you have taken into account all the features of your products. The clause helpfully contains a suggested list of 8 possible factors to consider as a starting point. Remember to include all the food safety aspects that pertain to the product such as pH, temperature control, available water, packaging, storage conditions and so on. It will be necessary to include any process variations, including packaging forms. For example a chilled product will have different associated hazards from a frozen product, a canned product will differ from an aseptic carton and so on. It sounds simple but it is worth spending time making sure that you do have all the relevant features written down: the Auditor's pen will start working fast if they see that you have missed a key feature of your product right at the outset.

The final bullet point is interesting. How do you predict how your customers might misuse a product? Here is one example.

Example

Evidently in the United Kingdom, a small number of consumers are known to graze on the toppings of frozen pizzas while they are defrosting, awaiting cooking. This strange behaviour causes some retailers concern.

Clause 2.3.2

This clause concerns the research you have put into your HACCP and in particular the background information necessary to it. When you first put your system together, you may well have included reference to the latest scientific literature, codes of practice, legislation and so forth relevant to the products. If you did not, it is certain a nonconformity would result. Be as wide-ranging as possible with your references to show a breadth of knowledge. The clause is helpful here in providing guidance with a list of six possible sources, including the very helpful 'customer requirements'.

Example

A sandwich manufacturer had prepared a reasonably good HACCP system but had made no reference to any guidelines or legislation.
   A nonconformity was given.

Note: The Requirement also states that your information sources be available on request. The Auditor may ask to see them.

A very important point here is that you must update this information. A key issue is legislation, which never stands still, and should be monitored for the latest updates.

> ## Examples
>
> A very good UK site producing dairy products had not kept their legal references up to date in their HACCP documents. In fact, they still referred to the Dairy Products (Hygiene Regulations) 1995 though these had been revoked by the Food Safety (General Food Hygiene) Regulations 2005. Also, they did not refer to any subsequent regulations such as the Food Safety (General Food Hygiene) Regulations 2006.
>   So they did not quite meet the requirement and it gave a feeling of uncertainty with the Auditor that they keep up to date with legislation, especially as it was specific to their industry.
>
> At a cheese company the hazard analysis included reference to food safety legislation but not to the latest legislation concerning allergens.
>
> A nonconformity was given in each case.

Clause 2.4

This concerns Codex Step 3: Identify Intended Use.

Key points in Clause 2.4
Who is the customer?
Vulnerable groups

Clause 2.4.1

This means 'use' by the consumer and is an important consideration that should not be overlooked. For instance, this must be considered where a product has to be prepared and cooked by the consumer, so perhaps preparation and cooking instructions need to be verified. As described in the clause, you must also consider the nature of the likely user and the suitability to vulnerable sections of the population. There are many aspects you might consider especially if you handle a diverse range of products.

> ## Example
>
> At a bakery for the potential vulnerable consumers assessment, in the HACCP study, did not consider allergen sufferers (e.g. nuts/sesame seeds/gluten in bakery products; and celery in prepared vegetable products), or infants (e.g. choking hazard of diced vegetables).
>   A nonconformity was given.

Where the ultimate consumer does not prepare the product themselves, for example if you are supplying a catering product it would be vital that you have considered the circumstances in which your products will be used. Similarly, your storage instructions should be considered especially for frozen or chilled foods.

The significant point here is that in such cases the Auditor may want to see evidence that you have considered hazards that might occur after the product has left your premises.

Clause 2.5

This concerns Codex Step 4: Construct a Flow Diagram.

Key points in Clause 2.5
Including everything
Risk zones

Clause 2.5.1

Everyone likes a flow diagram, especially Auditors. This should be no trouble to produce as it enhances everyone's awareness of your processes, including your own staff. It will also make a very useful aid to training new staff. Make the flow diagram as simple as possible and easy to follow but at the same time include all the necessary steps. It must be inclusive and portray the food processing within your HACCP scope. Do more than one if necessary where different processes are too divergent to encompass in a simple chart. You may need to consider steps both before and after the actual manufacturing process and any outsourced work as well. As a guide, the clause includes a helpful list of 9 possible points that should be considered.

Examples

At a bakery, there was no detailed HACCP study (particularly a flow diagram) for the repacking operation.

The flow chart for a tomato products canner dated 13/01/10 was incomplete (e.g. no inputs for herbs, tubes, steam, recycling of juice pasteuriser and waste output).

At a bakery, not all inputs are detailed on the HACCP process flow diagrams; notably divider / vegetable oil (dough handling surfaces), steam (plant bread oven baking), D-Tin (tin greasing agent).

At a tomato cannery, the flow diagram dated 01/09/11 showed no parameters for metal detector, wastes output from process and washing water chlorination.

HACCP process flow diagrams at a jam factory did not include reference to all inputs, for example steam, rework loops.

A nonconformity was given in all cases.

Note that one of the points is to show low-risk, high-care, high-risk segregation. This should not be confused with the requirements of Clause 4.3.1 that require a site plan to also show production zones. A flow diagram is, of course, a schematic illustration and not a representation of the actual layout. Remember also to include re-work:

Example

At a manufacturer of colours and flavours, rework is used for liquid and powder colours. However, the use of rework was not documented on the current flow diagrams for liquid and powder colours.
   A nonconformity resulted.

Note also that the flow diagram is part of the information that you must supply to the CB before the Audit (see Chapter 4) as it should give them a good insight and a headstart in understanding your processes for the audit.

Clause 2.6

This concerns Codex Step 5: On-site Confirmation of Flow Diagram.

Key points in Clause 2.6
Verifying and how often

Clause 2.6.1

The fifth step is to verify your flow diagram against the actual process. The Auditor will certainly have their minds on this as they go round the site themselves, so it is essential for you to carry out a check on all parts of the flow diagram. Codex suggests confirming the flow diagram during all stages and hours of production, so consider all shifts. In addition, any seasonal variations shall be considered.
   You must carry out your check by audit and challenge.
   Note: A change made for Issue 6 was that this challenge must take place at least annually.
   A further requirement is that you maintain records of verified flow diagrams. So, the Auditor will ask for evidence of internal audit and you should keep a history of any previous flow diagrams so they can see any changes made.
   It is sometimes easy to forget to update your flow diagram when you have modified a process. Make sure you keep it up to date.

Part Two

Part Two

Examples

A manufacturer of wet and dry food ingredients had not updated their flow diagram following a change to the spray drying process.

At a bakery, there was no evidence that the HACCP process flow diagrams have been verified as accurate during the previous 12 months.

A nonconformity was given in each case.

This is where we arrive at the Seven Principles of Codex Alimentarius. Clauses 2.7–2.13 follow these in logical order.

Clause 2.7

The three clauses under Clause 2.7 are concerned with Codex Alimentarius Step 6, Principle 1: List all potential hazards associated with each step, conduct a hazard analysis, and consider any measures to control identified hazards.

Key points in Clause 2.7
Allergen risks
Hazard analysis and what to include
Control measures and validating

Clause 2.7.1

There are several aspects to this clause. Firstly, you need to consider every process step from hazards present in raw materials as well as those from the process itself and those that might survive processing. Codex tells you to identify all your process steps, then list every possible hazard associated with each step and these should be classified as physical, chemical and microbiological, but remember to take into account allergens as well.

Example

A mushroom canner had some poorly defined analyses. Allergen risks, GM risks were not included, nor were condition of magnets.
   A nonconformity resulted.

The object is for you to be sensible here so that the system is focussed on genuinely likely issues. A document that lists everything under the sun is not likely to inspire the Auditor and in such a case there would be concern that proper logic had not been applied. However, you must be comprehensive in your considerations.

Examples

A milk dairy had an area where filling in glass was carried out. Clearly, glass itself is a hazard in such an area; however, the company had neglected to carry out an assessment of the risk of glass contamination here and had not listed it as a potential hazard.

Potential process delays at a bakery were not considered in the HACCP hazard analysis (e.g. bakery dough proved but delay in baking).

A nonconformity was given for both.

A key requirement is to record, that is document the potential hazards. As discussed in Clause 2.0, the Auditor needs to see the background to your system and what hazards perhaps were not considered later as requiring CCPs as well as those that were. It is an important point because if the Auditor can see what hazards you have considered they can see if you have been complete in your assessment or they may see that you have missed something important.

The position with allergens continues to be of great concern and it is very important that all aspects of allergens are considered here. The subject is covered in Clause 5.2. However, the Auditor will want to see evidence that allergens have been considered in this list, including any possible cross contamination that might occur, not only during processing but before and after, including raw materials.

Part Two

Example

A small company making bread rolls and other bakery products produced sesame seed buns. However, they had neglected to include the risk of cross contamination of sesame seeds into products that do not contain sesame as a potential hazard.

Therefore, they had not considered the risk of this potential allergen appearing in their sesame-free products and it was missing completely from their hazard analysis and their HACCP system.

A nonconformity resulted.

Note: You should consider foreign body hazards as part of your thoughts on physical hazards. At the same time in order to satisfy a later clause (Clause 4.10.1.1), you must consider potential equipment such as filters, sieves and magnets for the removal of foreign material and detection equipment such as metal detectors and X-ray.

Clause 2.7.2

This is a cornerstone of your HACCP: 'Conduct a Hazard Analysis'. You may have a smart HACCP plan ready to show to the Auditor but is this what the Auditor wants to see at this point? Most likely not. At this point, what the Auditor needs to see is the foundation work done in leading to a final HACCP plan. The Auditor needs to see what hazards have been considered, including all those that did not make it through to the final HACCP plan because they were not sufficiently significant. In other words, what the Auditor would like to do here is compare your original list of hazards (see Clause 2.7.1), with those that need to be prevented, eliminated or reduced to acceptable levels, that is CCPs.

So, to repeat, please keep all your original hazard analysis documents. They are sometimes discarded, which makes the Auditor's job difficult. If you have discarded them because the HACCP was put together some years ago, you should consider repeating the hazard analysis when you next review the system.

The clause lists seven factors that must be considered 'as a minimum'. You must therefore look at all seven and include consideration of these in your analysis; for example the first two: the likely occurrence of the hazard and the severity of the effects on consumer safety. The way in which you produced your list should firstly have taken into account their likely occurrence. Here, you should be realistic and rate the genuine chances of something actually happening. If the likelihood is so low as to be negligible, then you may need to consider it no further.

This should be balanced with consideration of the severity of the adverse effects on the consumer of something going wrong. In other words, if an event is unlikely yet the consequences of it going wrong are very severe then it should not be discarded. An example might be the survival of *Clostridium botulinum* spores in a canned product. Here, the chance of it happening in a properly run plant is low yet if it did happen the consequences could be literally fatal.

Examples

A produce supplier could not produce a complete hazard analysis on the day of the Audit and certain likely hazards had not been listed for example mycotoxins, pesticides, heavy metals, micro-organisms.

A manufacturer of cocoa liquor and cocoa butter had done a good job of deciding on CCPs and of educating staff into the importance of the system. Indeed, they had physically labelled each point of production with the CCP number. However, although they considered whether each possible hazard was likely to occur they had not considered the severity of effects.

A nonconformity was given in both cases.

It could be useful for the Auditor to be presented with a list or matrix of hazards with their likely occurrence and severity of effects rated. Limits can be set also. If the likely occurrence coupled with the severity exceeds the specified limit it becomes a risk where a CCP must be considered. In other words, based on those ratings, the significant hazards will be the ones that formed the basis of your final HACCP plan.

Furthermore, there is a practical reason for doing this when it comes to considering monitoring frequencies of CCPs later. If a hazard is considered significant such that it is a relatively likely event and the severity of adverse effects is high, then you will need to be monitoring this point quite often.

The vulnerability of consumers exposed to an issue is also something you must consider. As with Clause 2.4.1 where you considered the suitability of your products for certain vulnerable groups, here you must consider who might particularly suffer from a hazard and therefore what your acceptable levels of reduction might be.

Another item listed for consideration is the survival and multiplication of micro-organisms of concern. This should be a relatively straightforward point. The Auditor will want to see that you have considered all possible micro-organisms of concern to food safety and furthermore all the conditions that might affect their survival. For example these might include available water, temperature, pH and so on.

Example

A manufacturer of chocolate-coated fruit and nuts had not considered microbiological hazards at all in their HACCP study. In this case, the Auditor rightly knew that there are potential microbiological hazards that should be considered in chocolate manufacture, for example *Salmonella*.

A nonconformity was given against this clause.

Similarly, the presence of toxins, chemicals or foreign bodies must be considered. Toxins might include mycotoxins, which could be present in raw material on intake or develop through poor storage after intake. The presence of pesticide residues might also need to be taken into account.

Otherwise, the contamination of product at any stage, raw material through to finished product must also be included. Consider all possible sources of foreign bodies and therefore any means you require for prevention, such as sieves, or detection such as metal detection: refer also to Clause 4.10.1.1.

In addition to foreign bodies, toxins or chemicals you might also need to consider contamination by allergens, meat in vegetarian product, non-organic in organic, the presence of GM ingredients and so on. Remember, consideration of contamination of finished product might include attention to security or to packaging and how tamper-proof it is.

To repeat, there are six items listed as bullet points that must be considered in your Hazard Analysis and omitting any of them will result in a nonconformity.

Finally, where you have decided that a hazard cannot be eliminated but only reduced to acceptable levels in finished product, the clause requires that you justify these levels. In each case, you must document this. The Auditor will be particularly looking for such justification. For example this might be the case with certain microbiological limits.

Part Two

Clause 2.7.3

The final part of this Codex principle is to consider any measures needed to control the hazards you have identified. There are two aspects to be aware of here. The clause firstly refers to preventing, eliminating or reducing the hazards to acceptable levels and adds that you may consider more than one control measure. The Auditor will want to see evidence that you have done this.

This should lead you to your control points and ultimately your CCPs as discussed in the next clause.

Clause 2.8

There is just one sub-clause here and it is concerned with Codex Alimentarius Step 7, Principle 2: Determine the Critical Control Points.

Key points in Clause 2.8
CCPs and using a decision tree

Clause 2.8.1

So, this should be a straightforward one: 'Determine the Critical Control Points'.

I have mentioned previously that the Auditor will occasionally find that too many CCPs have been assigned, indeed sometimes a final HACCP plan looks like a list of every possible hazard and prerequisite that could be thought of and each has been labelled as a CCP. Sometimes, they are not even hazards at all in the true sense. Ironically, this seems to be particularly common in small companies with relatively simple processes.

Remember that a CCP is not the hazard, it is the point at which it is controlled.

Examples

A relatively large company specialising in preparing and packing fruits and vegetables had not considered prerequisite programmes when carrying out their HACCP study. At the same time, two 'CCPs' identified were really global issues and not related to a process step. In other words, they were prerequisites. The inclusion of these 'CCPs' appeared to be somewhat random.

A processor of rice and pulses had a system of sewing bulk bags of rice as a finished product. They had metal detection that they had identified as a CCP. In addition, they had identified the breakage of the sewing needle as a potential hazard (which was correct); however, they had also identified this stage as a CCP. This was incorrect because at this point there was no way in which to eliminate or reduce the risk.

At a tomato, cannery chlorination of cooling water, pH control and pasteurisation were not considered critical. Remember that the CCP is the point at which you control the hazard, which is not necessarily where the hazard occurs.

Therefore, a nonconformity was given in each case.

As long as the Auditor can see that all hazards have been identified and analysed initially, a good final HACCP plan should be a precise document, which enables the staff to focus on the real and likely issues rather than be a 'shopping list' of anything that can go wrong. For this, the CCPs should be determined in a logical way and, of course, both Codex and this clause points you in the direction of the Decision Tree process. Note though that Codex states that the decision tree approach may not work for all situations and that training in the application of the decision tree is recommended.

The Auditor will judge whether you have included all the CCPs that pertain to your processes so you need to be sure that you have included all your processes.

Example

A manufacturer of part-baked, fresh, frozen, filled, topped baguettes, garlic baguette slices, topped frozen bruschetta, filled, topped frozen ciabatta and so on and filled frozen sandwiches had no CCPs for the sandwich part of their production. The Auditor considered that this was an oversight.

Thus, a nonconformity was given.

It will be much easier for the Auditor to assess your procedure here if you have kept the records of your decision tree or other decision process. This can be your original records for each process step considered and may reflect the fact that your team was involved in making the decisions. These are often depicted in 'tree' form for each one. Alternatively, companies may include the decisions in short form among the columns of their final HACCP Plan.

Above all, it is important that the CCPs are clearly stated. In the following example there was confusion all round, which led to more than one nonconformity.

Example

A medium-sized dairy had a HACCP system that had been reviewed. They had a fully documented hazard analysis. However, it was not always clear that CCPs had been determined by the decision tree process in that some issues that should have been CCPs were not designated as such.

> HACCP Plans were in place, which set out to describe the process steps and associated hazards, the control measures, critical limits, monitoring procedures and frequencies, the corrective actions and responsibilities. So, the company had some idea of what was required, or at least how to set it out on paper. However, there was confusion where several issues had been combined together and it was difficult to match the appropriate monitoring requirement to the relevant issue.
>
> An example was under yogurt manufacture where it was unclear what the monitoring system for lactic acid level should be. Elsewhere, the terms 'limits' and 'monitoring' were clearly confused, 'hourly samples' being given as the limits in one case.
>
> In the event, they were given two nonconformities, one for failing to have clear CCPs and another for having confusing monitoring details (see Clause 2.10).

Sometimes, the Auditor may disagree with your decisions. While a good HACCP system should not be awash with too many random CCPs, it is important that the genuine ones are identified. Again, remember that the CCP refers to the process step, not the hazard itself.

> ## Example
>
> A manufacturer of chilled pies, quiches and puddings had several potential hazards to consider relating to the growth of bacteria due to inadequate temperature control. At one point, the hot holding temperature of gelatine was considered as a hazard. This gelatine was filled hot into cooked pies, which then had no further heat treatment. The company had not considered that this process point was a CCP. In the Auditor's opinion, there was no other point later in the process where this hazard could be reduced or eliminated, thus the Auditor considered that it should have been recognised as a CCP.
>
> Because this was one of several examples, a major nonconformity was given for this point.

Note that the final part of this clause is a reminder that where the control of a hazard does not exist at that point the product or process shall be modified either at the relevant step or elsewhere to provide control. If you have had to do this, retain the evidence.

Clause 2.9

All the clauses under Clause 2.9 are concerned with Codex Step 8, Principle 3: Establish Critical Limits for each CCP.

Key points in Clause 2.9
Setting critical limits
Making them measurable
CCP validation

Clause 2.9.1

This should also be relatively straightforward, but can be done incorrectly or occasionally overlooked altogether. The clause demands that you establish critical limits for your CCPs. In other words, as Codex suggests, you need to set and validate limits of tolerance for each CCP and this might include such parameters as pH, time, temperature, moisture content, available water (a_w), available chlorine and so on. Do not forget also that it could include sensory parameters such as appearance and texture, but remember that each critical limit must be measurable in some way.

The important point here is that the Auditor will want to see a clear description of what the critical limits are for *every* CCP listed.

Example

A manufacturer of high-risk products: ready meals and pies, needed to chill their cooked products as quickly as possible. They considered that the chilling of their products after cooking within a certain time was a CCP. They had not however documented critical limits for the time and temperature relationship for chilling of its cooked products. In other words, the HACCP plan did not describe the desired product chill temperature to be reached within a certain time after cooking (e.g. less than 5°C within 4 hours).

The Auditor gave a nonconformity on this issue.

In some cases, the critical limits are so obvious (or ought to be) that they should write themselves. Surprisingly, many sites trip up on this point.

Example

A manufacturer of cut, fresh, chilled poultry packed the products under modified atmosphere. This process step was identified as a CCP, and the manufacturer tested the packs for residual oxygen. However, the Auditor found that the maximum acceptable critical limits for residual O_2 levels had not been identified or documented.

A nonconformity was given.

Other situations are perhaps less clearly defined, and some companies seem to have difficulty in defining critical limits for them. So they ignore the concept altogether, leaving the section on critical limits in their HACCP plan deficient.

Part Two

> **Example**
>
> A bakery producing cakes, pastries and other confectionery had the sieving of flour down as a CCP for preventing contamination. However, they had not given a critical limit for this, such as a mesh size for the sieve.
>   A nonconformity was given.

Another important point is that you should also make sure that there has been no drift away from your critical limits by operatives using different parameters. This kind of thing should be picked up on internal audit if not before.

> **Example**
>
> A cutter and packer of cheese products had metal detection as a CCP and had listed test piece sizes as critical limits, so all should have been well and good. However, the test piece sizes being used on the production line were different to those written down.
>   A nonconformity was given.

Referring to any critical limits that are based on subjective data such as visual inspection you must show clear guidance or examples to satisfy this clause. This might include photographic standards, colour charts or written descriptions. The Auditor will certainly pick up on any such critical limits and ask to see such guidance, so you should have these available.

Some canneries have successfully used visual seam checks as a CCP (see Clause 2.10.1). In such a case, there must be full instructions and guidance on this control.

Clause 2.9.2

This clause requires the validation of each CCP such that you can demonstrate that the control measures in place are consistently controlling the hazard to within the critical limits. You must have documented evidence of this.

The word 'consistently' is important here. If you are getting the odd 'blip', you may need to review your control measures. Worse still, if you are repeatedly failing to meet critical limits, there is clearly an issue with your control measures.

The Auditor will want to see positive evidence that you are carrying out such validation. What they will be looking for is a system of checks, audits and reviews. This can be done by daily checks on the records of CCPs or by auditing of the system or a combination of both. You should also consider inputs from outside such as customer complaints. The Auditor will want to see records of checks and audit reports and will be particularly interested in any nonconformance reports (see Clause 2.13). These will show that you are on the ball and show how you deal with such events.

Examples

A manufacturer of bread rolls had not validated their HACCP system by any means since first putting the system in place. Furthermore, that had not audited the HACCP system properly such that the operation of CCPs was checked.

At a bakery, there was no documented justification as to the defined critical limits for CCP3 (metal detection test-piece sizes).

A nonconformity was given in each case.

Clause 2.10

All the clauses under Clause 2.10 are concerned with Codex Alimentarius Step 9, Principle 4, which states: 'Establish a Monitoring System for Each CCP'.

Key points in Clause 2.10
Monitoring each CCP
Rapid response
Ensuring records are complete and verified

Clause 2.10.1

This is extremely important. You must have a system of monitoring your CCPs and they must be well documented and the results well recorded. The Auditor will want to see a brief synopsis of the monitoring of each CCP in the HACCP plan, usually backed up by more detailed procedures. Note also that later in the Standard, in Clause 7 concerning staff training, there is a requirement (Clause 7.1.2) that all personnel involved in activities relating to CCPs must receive *relevant* training. So, your procedures here will be an essential part of that training also.

The Auditor will be particularly interested in this information because they will definitely use some or all of your CCP monitoring procedures to audit during the site tour and subsequent document review. So, if they have not been documented properly, or are not being followed properly you will be given nonconformities. Your records will be key and they must show positive conformity as well as any issues. Monitoring procedures must be followed.

Examples

At a bakery, there were no positive CCP1 monitoring records maintained for the visual appearance of in-coming goods (only a comments section when nonconforming materials are received).

> A manufacturer of ice lollies, using three daily shifts, had identified filtration as a CCP and had set out to check the filters every shift as their monitoring procedure. The Auditor noted this but when looking to see what was actually happening on the production line, found that the operatives were only checking on a daily basis.
>
> At a bakery, a CCP (chilled/frozen storage temperatures) was not being monitored using a calibrated hand-held temperature probe (as required in the CCP procedure) but by reading off from the chiller's uncalibrated built-in temperature gauges.
>
> Nonconformities were given against this clause.

As with the critical limits, it is important to be accurate when describing the monitoring procedures, giving the method of monitoring, the frequency and an indication of those responsible for carrying it out. There is no necessity to have too much detail on the HACCP plan here as this can be covered properly in a detailed procedure document. In this case, the procedure should be referenced here. Remember that monitoring might include laboratory testing, so your testing methods should be referred to as well as the frequency of testing.

The monitoring procedures must be adequate in that they must be able to detect loss of control. The Auditor will be looking for CCPs to be in control and for a system that adequately monitors this.

Example

A company involved in the preparation of cured and cooked meats had the cooler exit as a CCP with the exit temperature as a critical limit. However, the Auditor noted that they were measuring temperature at only one point of the cooling rack on exit. In the Auditor's opinion, temperatures should have been taken at the top, bottom and middle of the rack because of the potential for temperature variation.

Hence, a nonconformity was given.

The adequacy of a monitoring system is also determined by the ability to provide information in time for corrective action to be taken and this too is a requirement of this clause. For instance, in the canning industry, more reliance has come to be placed on the visual inspection of can seams in recent times as an effective way of monitoring the seams because it can be done much more frequently and rapidly than the complete 'tear down' check and thus corrections can be made more readily. Full seam checks must also be done, but the visual check can be a very good way of identifying a problem earlier. Remember that such visual checks must be supported by clear guidance on critical limits (see Clause 2.9.1).

As a guide, the clause lists four points to consider concerning the types of measurement you might take. The examples given previously are of discontinuous measurement but an important tool in many situations is continuous measurement for example of temperature. In some industries such as the pasteurising of milk or in canning, continuous temperature measurement is essential.

Clause 2.10.2

This is an important clause concerning records but one that should be easy to comply with. All CCP monitoring records must include the results of the measurements, be dated, signed by the person responsible for the monitoring and verified by an authorised person. This might seem obvious but evidently not to all.

Example

A produce packer had CCP records that were unclear. There was no distinction between CCPs or 'QCPs', they had not been properly verified.
    A nonconformity was given.

The Auditor is likely to spend some time going through records of CCP monitoring, both from your archives and during the site tour. This is one of the clauses in the Standard that is particularly exacting in its requirements on records. So, you should ensure that all your paperwork is in good order, dated and signed.

Examples

At a bakery, there was no evidence of an authorised person verifying the CCP monitoring records

At a mushroom cannery, it was observed that registration of seaming control on line 3 and line 5 (CCP3) is conducted on the same form. It would be difficult to identify which line was out of control.

At another bakery, CCP3 (yeast receiving filter) records from March to September were not verified by an authorised person.

A nonconformity was given in all cases.

Note: (Introduced for Issue 6) If your records are electronic, you will have to demonstrate 'signing' and verification by means such as restricted login or access.

Clause 2.11

All the clauses under Clause 2.11 are concerned with Codex Alimentarius Step 10, Principle 5, which states: Establish the corrective action to be taken when monitoring indicates that a particular CCP is not under control. Here, the Standard refers to a Corrective Action Plan.

Key points in Clause 2.11
Making sure action is definitive

Clause 2.11.1

Once you have established your critical limits and how you are going to monitor them, you must then state exactly what your corrective action is when a CCP is found to be not within critical limits. Note that the clause requires action also when results indicate a trend towards loss of control.

This has to be definitive action, documented in the HACCP plan and you must nominate personnel to take such action. The Auditor will not be satisfied with, for example a statement like 'report failure to the engineer' when a critical piece of equipment has failed. What the Auditor will want to see is proper description of action to be taken to bring the CCP back under control.

This must also include the action taken on product made while the CCP was not under control. For example, you will need to isolate any nonconforming product and decide on its fate.

> ## Examples
>
> An ice cream manufacturer had raw materials in chill storage while finished product was held in cold storage. While they monitored temperatures of the two areas and had set tolerances, they had not set down what corrective action to take should the storage temperatures fall outside of those limits. For example, nothing was said about what to do if the freezer had failed for several hours.
>
> At an automated bread bakery, the Auditor found that the corrective action for when the freezer temperature was outside of set limits just stated 'call engineer'. Also, there was no indication as to what to do with affected product.
>
> A nonconformity was given in each case.

Clause 2.12

Clause 2.12 is concerned with Codex Alimentarius Step 11, Principle 6, which states: Establish procedures for verification to confirm that the HACCP system is working effectively.

Key points in Clause 2.12
Verification of CCPs and prerequisite programmes
Recording and communicating results

Clause 2.12.1

This clause has been merged from two clauses from Issue 5. The first part requires you to establish procedures of verification to confirm that your HACCP system is working effectively. This means looking at your hazard analysis, your CCPs, your monitoring systems and so on and making sure that they are still correct and that the monitoring systems work. This is just like verifying a method. In other words, you make sure that, when followed correctly, your systems are sufficient and effective. For this, you will be taking a global view of your HACCP and, very importantly, prerequisite programmes. The clause gives four examples of how this might be done practically, which include internal audits and the review of various sources of incidents.

What the Auditor will be looking for here is a system of checks, audits and reviews. For the Auditor to see that you can verify that the HACCP plan is effective, you will need evidence, for instance that you have checks on CCPs being monitored and recorded and that your prerequisite programmes are reviewed. This can be done by daily checks on the records of CCPs or by auditing of the system or a combination of both. The importance of audits cannot be stressed enough. The Auditor will want to see records of checks and audit reports and will be particularly interested in any nonconformance reports. These will show that you are on the ball and show how you deal with such events.

Examples

At a manufacturer of sweetener sachets, the Auditor found that the information gained from their HACCP verification audit did not match daily records. For example, metal detection testing was an hourly requirement. Records indicated that testing as actually done twice daily instead of hourly. This was not picked up on their HACCP audit.

A manufacturer of dried potato granules, which was part of a very large group had their system verified by their central technical function. However, they had sieving as one of their CCPs and had not verified that this was being monitored properly by audit or any other means.

A nonconformity was given in each case.

You should also consider input from outside the company such as customer complaints, incidents with local authorities and withdrawals or recalls. Any such issues should reflect either on CCPs or prerequisite programmes.

Secondly, your verification results must be recorded and the results communicated to the HACCP Team. The Auditor will want evidence of this: the records themselves and evidence that the team have been made aware such as memos or minutes of meetings.

Clause 2.13

The single clause under Clause 2.13 concerns the final Codex Alimentarius Step 12, Principle 7, which states: Establish documentation concerning all procedures and records appropriate to these principles and their application.

Key points in Clause 2.13
Records of CCPs and prerequisite programmes

Clause 2.13.1

This clause is all about documents and records. I would hope that by now you have got the message of the importance of keeping good records and of having them available. In this case, the clause requires that all documentation (which would include all plans and procedures) and records be sufficient to assist you in verifying that the HACCP is working and that controls are in place. Note that for Issue 6 you must also include controls managed by prerequisite programmes. This means that you will need good records to show conformity as well as nonconformity.

Of course, you can only maintain records that you have created them properly in the first place.

> ### Example
>
> A cereal products manufacturer had identified the possible presence of ergot in incoming grains of wheat as a CCP. However, there was no positive record for maintaining control, that is for the absence of ergot in incoming grain.
>
> Therefore, a nonconformity was given on this point.
>
> The same factory had also identified the presence of congealed lumps in finished product as a hazard. In this case, records of the presence of these lumps were not being kept. Consequently, they were not recording nonconformity at this point and had again failed to satisfy this clause.
>
> Therefore, a second nonconformity was given for this clause.

Clause 2.14

This final clause in this section concerns the very important subject of review.

Key points in Clause 2.14
What to review
When to review
Prerequisites

Clause 2.14.1

Essentially, the HACCP plan must be reviewed at least annually and prior to any changes taking place that affect product safety. Note that for Issue 6 this review shall also include prerequisite programmes.

> ### Example
>
> A factory making water ice lollies had a good documented HACCP system, which met most of the requirements. However, there was no evidence to show that the system had been

reviewed. There were no issue numbers on the documents, which might have indicated a review process and no other review records.
    A nonconformity was given.

Regarding changes, as a guide, examples of such changes are listed in 7 bullet points in the clause and include both internal and external changes. Note that a new one for Issue 6 is: 'emergence of a new risk, for example adulteration of an ingredient'. The Auditor will be looking for evidence that you are keeping up with change and for example might ask you if there have been changes to staffing or management. You might also be asked if you are aware of any relevant changes in legislation, for instance. The Auditor may then check to see if known changes are reflected in HACCP and prerequisite programmes review.

Naturally, any new products should trigger a review. This is a point that companies sometimes overlook and forget to update systems when new product types have been added to the range.

Example

A manufacturer of dried and blended food ingredients also produced oil and fat-based pastes and some agglomerated products. The agglomeration process was carried out on product already dried in an earlier part of the process. While they had a fully documented HACCP system, which included the drying process and the production of the pastes, they had not considered the agglomeration process.
    Therefore, a nonconformity was given on this point.

The review must be carried out by the HACCP team. If the Auditor feels that the review is lacking in this way, it may lead to a nonconformity being raised.

Example

A grader and packer of eggs had carried out a HACCP review some months before the audit, but it appeared to have been a one-person effort. The company had a HACCP team but there was no evidence that it had been involved in the review. The Auditor considered that the team should have been involved in the review.
    A nonconformity was given.

Summary

HACCP is ultimately about logic and a strong foundation in HACCP will pay dividends. Focus on having a good team and a broad analysis of all possible hazards.

Part Two

Make sure your documentation is good throughout including records. Get this right and you are sure to do well.

Quiz No. 7

(1) How many clauses in this section mention prerequisite programmes?
(2) How often must you verify your flow diagrams?
(3) Which clause requires that you consider customer misuse of a product?
(4) Does a clause require the HACCP Team Leader to have training?
(5) Critical limits shall be measurable wherever possible and what else?
(6) How often must you review HACCP?

13 Clause 3: Food Safety and Quality Management System

Clause 3 is largely concerned with systems and procedures and as such almost every aspect of this section requires you to have documentation and records to be able to demonstrate compliance. The clauses are predominantly green colour-coded. Note that of the ten Fundamental Requirements, three are within this clause (Clauses 3.4, 3.7 and 3.9). This illustrates the need to have a significant depth of records and evidence to meet the requirements of this section. Overall, Clause 3 is likely to take up a relatively large part of the Auditor's time in following audit trails and tracking down and examining documents. It is also likely to be a major part of the Part 2 audit where an unannounced audit is taking place.

In previous issues, this section began with a reference to a Food Safety and Quality Policy. This subject now appears in Clause 1.

Clause 3.1: Food Safety and Quality Manual

Key points in Clause 3.1
Quality Manual and procedures
Ensuring it is implemented
Making it available
Detailed, legible and in different languages

Here, we come to another one of the cornerstones of your system. With this series of clauses you are required to have a Food Safety and Quality Manual including procedures. The SOI sets out the aims and objectives of having documented procedures, which are:

- to facilitate training;
- to support due diligence; and
- the production of a safe product.

The BRC Global Standard for Food Safety: A Guide to a Successful Audit, Second Edition. Ron Kill
© 2012 John Wiley & Sons, Ltd. Published 2012 by John Wiley & Sons, Ltd.

The Manual must be the embodiment of your commitment to food safety and quality, citing training and due diligence specifically. For those who are not familiar with the phrase 'due diligence', it relates to a company's potential defence under UK law when defending against legal action in a food safety case. In the United Kingdom, there is strict liability in food safety cases, which in effect means that when such a case is brought, it is up to the defence to prove innocence. In such cases, the defendant must try to show that due diligence and all reasonable care was taken in the manufacture of the product. Of course, once you are in a courtroom, documented proof will be necessary, hence the reason it is so important to keep documented procedures and records against the day this happens.

Clause 3.1.1

Food Safety and Quality Manuals come in many shapes, sizes and formats. What the Auditor is looking for is a coherent Manual that ties together policies, procedures and work instructions that you have issued to your staff. It should be indexed. What the Auditor does not want to see is a random collection of procedures, lacking in document control, all bound up in one ring binder. This is not what is meant by a Food Safety and Quality Manual, but sadly it has often been the case in the past that the company proudly presents the Auditor with such a document.

Example

A large milk dairy had a Quality Manual and documented procedures. However, the Auditor felt that the system was disjointed. The Quality Manual did not have references to the related procedures. Such references would have demonstrated tighter control.
   A nonconformity was given.

Unlike HACCP, there is no requirement for the Food Safety and Quality Manual to be written by a team. Generally, they are written by a Technical or Quality Manager, sometimes by a consultant. The Food Safety and Quality Manual may well include your HACCP system, but if not, it should make reference to it. In addition, it is likely to include some or all of the following processes:

Food Safety and Quality Policy (ref Clause 1.1.1)
Organisational structure and responsibilities
Management commitment
Management review
Resource management
Internal audit
Purchasing
Supplier approval
Documentation control

Specifications
Procedures
Record keeping
Corrective and preventive action
Traceability
Product withdrawal and recall
Complaints
Maintenance
Hygiene
Waste
Pest control
Product control
Allergen control
Control of nonconforming product
Process control
Personnel

As the Standard itself might say, this list is not exhaustive.

The Auditor may carry out some audit trails during your audit starting with a procedure, then work instruction (if separate), taking note of any relevant forms or documents referred to. If procedures are not being adhered to it, will lead to nonconformities being raised.

Finally, for Issue 6, the Standard mentions that your Manual may be in printed or electronic form.

Clause 3.1.2

This clause requires firstly the Food Safety and Quality Manual to be fully implemented. This is a good example of a requirement that is in effect saying that you must have had sufficient time to develop your systems and have them all in place and working. It means that every aspect of the Manual shall have been issued and be adhered to.

The Auditor is likely to ask the simple question as to whether the Manual is fully implemented, but will also probably sample certain sections and audit them on site to be satisfied that it is so. Management reviews and amendments to the Manual will give the Auditor an insight into your review process.

This clause also requires you to make it available to all key staff. It is still sometimes the case that a Quality Manual is 'owned' by the Technical Manager to such an extent that only one hard copy exists and some key members of staff do not have easy access to it.

Clearly, one hard copy may be sufficient if others have access to it. Most conveniently, Quality Manuals that exist in electronic form are easily made accessible to others on a site via a computer network.

Part Two

Clause 3.1.3

Here, the requirement is essentially for the procedures and work instructions to be well written, legible and with sufficient detail for the persons who have to apply them. There is also a requirement for them to be accessible, thus they may also have to be in more than one language.

The Auditor may compare your policies and procedures with what work instructions or records sheets you issue to operatives. In the case of a CCP, this is particularly important.

Example

A manufacturer of frozen Chinese-style products provides an example. Here, the problem was that they had set down delivery temperatures of raw materials as a CCP but the critical limits or standards for temperature were not included on the goods-in check sheet where temperature was recorded. Given that the personnel here did not have easy access to the HACCP plan, it was considered that the temperature standards should have been included on the form. Furthermore, there was no procedure to follow should product be outside specification on intake.

A nonconformity was raised.

In a change for Issue 6, you must also think about using pictorial information in work instructions if it helps with conveying the information. As written in the clause, this may be necessary where you have many languages on site, cases of dyslexia or indeed illiteracy. Remember that dyslexia and illiteracy are emotive subjects and staff may conceal problems that they have with reading. It is something that companies need to be aware of.

Clause 3.2: Documentation Control

Key points in Clause 3.2
Effective control of documents
Make a list of documents
Method of control
Reasons for changes
Replacement of documents

This SOI is an objective for an effective system of control for all documents. Here, the Auditor will be looking at all documents starting with your Food Safety and Quality Policy, Quality and Procedures Manuals, your work instructions, record sheets and forms right throughout your process. It will be expected that all these are controlled and be the correct version. As a minimum, they should either be dated or bear an issue number; having both is not a bad idea.

> ### Example
>
> A large-scale liquid milk dairy had some elements of document control in place but key documents were not included within the document control system, notably final product weight control guide and cap torque test record.
>   This resulted in nonconformity.

Of course, for the Auditor it is also about the evidence that documents provide for the purposes of their audit. It should be added that throughout this clause and sub-clause, documentation can mean paper (hard copy) or electronic form (sometimes called 'soft copy'). The same rules will apply to both.

It is often the case now that soft copy versions of Food Safety and Quality Manuals and procedures are issued to staff. In such cases, the Auditor will seek assurance that there are means of protecting such documents from modification by unauthorised personnel. The Auditor will also ask what backing up facilities and procedures are in place to protect the system as a whole.

You might think that certain documents do not need to be controlled but if in doubt it is better to be cautious and control them anyway. Remember too that if later you are correcting nonconformities from the audit by amending documents (see Chapter 19), any corrected documents should be properly authorised re-issues. The CB will be looking closely at dates on any corrective action documents to be sure that amendments have indeed been made and dated correctly.

The Auditor may ask to see examples of 'masters' to be sure that the correct versions are in place on site.

The Auditor is likely to take notes of issue numbers of documents from your document list (see Clause 3.2.1) and compare them with those in use. Remember to consider all such documents. Here are some examples to show the range that you must consider.

Part Two

> ### Examples
>
> A situation was noted at a manufacturer of frozen and fresh poultry portions whereby HACCP documents were not authorised or dated, also the complaints procedure in place was not the correct issue.
>
> A large cider factory had some good document control but out of date versions of procedures being used in pulp/press room area, dated 3–5 years earlier, were noted. They had since been updated on the electronic system but clearly not issued properly.
>
> At a manufacturer of meat products and black pudding, it was noted that the black pudding recipe in use was dated 16/09/04 not 12/8/05 as amended.

A modern bakery had an out of date preventative maintenance record in use that did not match the current version as issued.

It was noted at a produce manufacturer that both the vacuum pack seal record sheet and weight control record sheet had not been included within the document control system.

All these resulted in nonconformities being raised.

It can be seen that to issue properly authorised, amended master documents is one thing but to ensure that the current version is being used in every corner of your factory may be another. Clause 3.2.1 concerns the replacement of documents but it is also recommended that this aspect of document control also be part of your internal auditing.

Clause 3.2.1

This clause concerns the procedure for managing documents. Note that this itself will need to be documented (and therefore controlled). The clause then lists four bullet points, which this procedure must include. The first two are newly stated in Issue 6 but are hardly new ideas. You must have a list of all controlled documents and indicate the latest issue number. At the same time, you must state the method of identification of documents and how they are authorised. Part of a good document control system is to record the reasons for changes to documents. This clause requires you to do so. It is a simple requirement and the Auditor will be looking for either a log of changes, giving the reasons or, as in some cases, reasons for amendment recorded on the actual document itself. The latter sometimes appears in the footer. Both methods are acceptable.

It is surprising how often this simple requirement is not met.

Examples

A processor of frozen sweetcorn had recorded the reasons for changes to HACCP documents but reasons for changing other critical documents were not recorded, for example for the Quality Manual or Procedures Manual.

At a bread bakery, it was noted that the reason for changing flour receipt procedure was not recorded.

At a different bakery, documented bakery recipes were not suitably controlled/formalised with numerous handwritten changes to ingredient weights and some totally handwritten recipes on uncontrolled paperwork.

Nonconformities were given in each case.

The final point concerns a system for ensuring the correct replacement of documents as they are updated.

Example

A processor of sweetcorn had no effective system for recall and replacement of key documents.
A nonconformity was given against this clause.

Clause 3.3: Record Completion and Maintenance

Key points in Clause 3.3
Genuine, legible, retrievable
Making alterations
Back up electronic records
Defined period
Shelf life

Hopefully by now you will have already got the message that record keeping is a vital part of your systems. This SOI quite simply requires that you keep genuine records to demonstrate control. Of course, this is a necessity and the Auditor will need to see many records of all aspects of your systems. As always, if you are maintaining computer records, the Auditor may ask about security, backing up and so on. Records must be genuine.

Example

At a dry ingredients supplier, it was found that finished product inspection records in the blending plant were pre-signed by the operative.
A nonconformity was raised.

Clause 3.3.1

This clause has several points to make about your records. That they shall be legible and genuine requires no explanation but will require some judgment on the part of the Auditor. The Auditor is likely to look at a broad sweep of records. Badly presented records with excessive alterations especially with correction fluid will not give a good impression.

Records must also be in good condition and retrievable. The Auditor might challenge this by asking to see a record from the archives. There is nothing wrong in occasionally having to alter a record. However, when it is done it should always be in an authorised and controlled way.

> **Example**
>
> At a bakery, unauthorised alterations were noted on metal detection records for June and August.
>   A nonconformity was given.

The Auditor will be reasonably happy to see the occasional alteration in pen with an initial and date by the side. Layers of correction fluid with no way of seeing any original work or who has made the changes will not be acceptable.

> **Example**
>
> When the Auditor examined records at a distillery it was found that there were several instances of alterations to records that were not authorised. They were sometimes crossings out but there was also liberal use of correction fluid.
>   A nonconformity was given.

Note: Correction fluid is an unsuitable material to have in a production area anyway. Finally, as with other documents, all electronic records must be suitably backed up.

Clause 3.3.2

The first part of the clause is straightforward: you shall have a defined retention period for records that relate to the shelf life of your products. As it must be defined, the Auditor would expect to see this in the procedures. The Auditor will establish how long you keep your records for; let us say, it is 3 years. The Auditor might then challenge this by asking you to produce a sample of records from that period, so be ready to produce something from up to the full extent of your retention period.

The clause also requires that any legal or customer requirements are taken into account when deciding how long you retain your records. One example of this would be where the legal process can take a certain time. You are likely to have to keep records of quantity checks for at least 12 months by law. As a result, the clause requires that you keep them for the length of the product shelf life *plus* 12 months. Naturally, your customers may also have their own requirements as to what you keep and for how long. Some standards, for example Halal, may have special requirements on the retention of records.

In the United Kingdom, a local authority has up to 1 year in which to take action on a complaint against a product. Therefore, it is advisable anyway to have a minimum period for the retention of records of the shelf life plus 1 year. Most companies seem to have little difficulty in retaining records for 2 years, thus any product with a shelf life of less than 1 year could be covered by this. Clearly, for products with very long shelf lives such as canned goods, the period will be longer.

This much is straightforward for products where there is no possibility of customers extending product life, for example by freezing. However, the clause then asks you to consider this point also in determining your document retention period. Certainly, where a product bears the legend 'suitable for home freezing' this must come into consideration, but there may be other products that consumers routinely freeze that may not be marked with this information. Bread is a good example.

It is not possible to predict how long the odd customer will keep a product for in their freezer, so the Auditor should use common sense here. They will be looking for a sensible and reasonable period of retention. Recently, a manager of a bread bakery telephoned me to ask about this and said that as his products had a shelf life of only 8 days did he need to keep his records, such as recipe sheets for any longer than that. As explained, the answer was 'yes'.

The Auditor will expect to see a defined retention period that complies with this principle.

Examples

A manufacturer of food ingredients and flavourings was retaining records for reasonable periods but had not defined a specific period anywhere in their systems and procedures.

At a frozen poultry producer, quality records were being stored for 18 months, which is an equal duration to the shelf life of the frozen product.

At a wine supplier, the Auditor considered the record retention time of 2 years to be insufficient given that the product could easily be matured by the customer for a longer period.

Nonconformities were given in each case.

Part Two

Clause 3.4: Internal Audit (Fundamental)

We now come to another cornerstone of your systems, the requirement for internal audit. This SOI is a *Fundamental Requirement* and rightly so, for there is a need to show a good track record over time. The objective is to verify that you are effectively applying your food safety plan and thus can meet the Requirements of the Standard.

As mentioned in Chapter 4, the BRC have published a guideline devoted to this subject that illustrates why it is important to audit, suggests audit frequencies and how you might report your auditing.

Key points in Clause 3.4
Risk-based programme
Annual schedule
Trained, independent auditors
Implemented procedures
Conformity reporting
Timescales for corrective action
Hygiene and fabric inspections

Many are the times, when Auditors have asked to see the internal auditing system in a factory and a company has proudly pulled out a factory hygiene audit system, expecting this alone to satisfy this clause. So, be clear that this objective is *not* just about hygiene audits, or rather it is about much more than hygiene audits. It is about the auditing of those systems and procedures that cover the requirements of the Global Standard and should include all systems within your FSQMS and HACCP that are critical to product safety, legality and quality. For instance, you would probably include:

- all CCPs,
- all pre-requisites,
- quantity control,
- QA and/or QC functions,
- customer complaints,
- management review, and
- nonconforming products.

There may be others but this is just to give an idea of the kind of scope you should be thinking of. The simplest way to cover this may well be to audit every procedure that you have to see if it is being operated according to those procedures.

Internal audit is such a good management tool that it is always a surprise that some companies do not take it more seriously. Being a Fundamental Requirement, a failure to comply with this SOI could result in a failure to achieve a certificate.

Example

A cutter and packer of meats and poultry, which was part of a group, had lost several key members of staff since their previous audit. The group had placed a new Operations Manager in charge, who also had responsibility for quality and technical management. However, he had had no proper training in the requirements of the FSQMS, being essentially a butcher by training. When they had their audit some months after his appointment, many of their quality systems were dormant, including internal audit, which had not been done for over a year.

The company's current certificate was suspended while a Grade D (this would now be 'no grade') was given for this audit and no new certificate was achieved until after a further, more successful audit took place some time later.

Clause 3.4.1

This clause is firstly a requirement to plan your audits. There is no substitute for a simple table or matrix here. The Auditor will clearly be able to see what you have scheduled and be able to consider whether your audit plan is sufficiently comprehensive. The Auditor might then ask to see the reports of some of the audits that have already taken place.

Example

A manufacturer of fresh meats was carrying out some internal auditing but there was no plan or schedule at all.
   A nonconformity was given.

The clause also asks that you set the frequency for each audit according to the perceived risks associated with each activity, and previous audit performance. The maximum interval between audits is 1 year. Of course, you may do certain aspects more frequently if your risk analysis dictates this or if previous audit performance has been poor. The Auditor will want to see that you have considered this in some way and will ask you how you arrived at certain audit frequencies. If you have had certain problems, for example an increased complaint level or a high number of nonconformities in a certain area, but have not increased audit frequency in this area, the Auditor would question this.

Part Two

Example

A liquid milk dairy had an audit schedule that it was meeting. However, there was no evidence that the company had considered the risks associated with each activity in setting their schedule.
   A nonconformity was given.

A good way to approach this might be to rate each aspect for audit according to the inherent risk of the activity, including any issues or nonconformities that have arisen and set the audit frequencies accordingly. You could also include the audit frequencies in your management review.

Clause 3.4.2

This is a two-part clause. Firstly, the requirement is for the internal auditors to be appropriately trained and competent. It is important for you to know what is meant by this. Auditing is a skill and is certainly one that a person gets better at with time and experience. However, it is also preferable to have some kind of formal training, as there are certain techniques that are difficult to pick up by oneself.

Therefore, it is probably a good idea for at least one person, for example the Technical Manager, to have had some formal external training in auditing techniques. The Auditor will ask about training and want to see any evidence of such training (e.g. certificates).

You will certainly need more than one internal auditor and the Auditor will take a view on the degree of training needed for any other internal auditors that you use. This may depend on the amount of auditing the other staff are doing. For example, if you have a well-trained Technical Manager who is doing 90% of the internal auditing, it may be that the others involved will need less training and it may suffice to have documented 'in-house' training.

> ### Examples
>
> A relatively small packer and grader of produce had a system for internal audit; however, none of their staff had any training in auditing. The Auditor felt that this was necessary for at least one of their staff.
>
> At a produce packer, there was no evidence of either training or qualification for internal auditors.
>
> Nonconformities were given.

Even if you have trained auditors, remember that the Auditor needs to see hard evidence that this is the case.

> ### Example
>
> A canner of peaches and apricots stated verbally that their main internal auditor had formal training in auditing techniques. However, they were unable to produce any evidence of this, such as a certificate, during the audit.
>   A nonconformity was given.

Always remember that you may have to update training as personnel move on.

> ### Example
>
> A baker was found to have no internal auditor training for the new Technical Manager (main internal auditor on site).
>   A nonconformity was given.

As regards competency, the Auditor is likely to judge this on the evidence of the audit reports.

The second part of this clause is equally important. The internal auditors must be independent of what they audit; in other words, they must not audit their own activities. Problems will occur here when a company does not have the resources for staff to be sufficiently trained in auditing and the Technical Manager for instance, attempts to audit everything.

> ### Examples
>
> A large processor of oils and fats had a Technical Manager who was carrying out all the internal audits, including activities within the technical department. Therefore, this could not be considered to be completely independent.
>
> At an ingredients supplier, the internal audit of the HACCP system was carried out by the HACCP Team Leader, who was, therefore, not sufficiently independent of the area being audited.
>
> A nonconformity was given in each case.

Clause 3.4.3

This clause arises from the merger of three clauses in Issue 5. Firstly, the requirement is for the programme to be implemented. Once you set your schedule, you have to meet it.

> ### Examples
>
> A manufacturer of canned tomatoes, pulses and fruits had an audit schedule for the year but had not managed to get through it all.
>
> A poultry processor had an audit schedule that determined that all aspects were to be audited half-yearly. However, this appeared to have been neglected as each area was being audited annually.

Part Two

> A frozen seafood company had not managed to audit their maintenance systems or complete their supplier audit programme according to their own schedule.
>
> A nonconformity was given in each of the three aforementioned examples.

Next, it concerns reporting and if you have read so far then you will already be well aware that the Auditor needs to see records to be satisfied that you have met many of the clauses. In this case, there are two specific points to make.

Firstly, you need to show evidence that you have identified conformity when carrying out your internal audits. In other words, your audit reports should not be simply a list of any nonconformities that have been identified, a kind of reporting by exception. Your reports should show a positive confirmation that you have been through each point.

Examples

A tomato cannery had carried out internal audits but had only reported nonconformities. Thus, they had not met the requirement and the Auditor could not be sure about the thoroughness of their audits.

At a bakery, the internal audits carried out revealed that there was no procedure in place to record conformity when carrying out internal audits.

A nonconformity was raised in each case.

Next, you must record nonconformities, of course. Most people make some kind of record of nonconformance but they are not always as full and clear as they should be.

Example

A manufacturer of frozen Chinese-style products had an internal audit schedule that was being adhered to; however, their internal audits raised nonconformances as a comment only and in many instances did not result in a system of corrective actions that were documented or signed off.

A nonconformity was given.

For guidance, you should consult Global Standard Guideline on internal audit. What is needed is clear reporting of the nonconformities and evidence that the corrective actions required have been brought to the attention of the appropriate personnel. Timescales for implementing them must be agreed with those personnel.

Examples

A manufacturer of cakes and snack bars had carried out internal audits but the responsibilities, corrective action timescales and verification of internal audit nonconformities were not formally assigned and documented.

At a supplier of processed peas, there was no indication on several of the internal audit reports that corrective actions had been carried out or verified by the auditor.

A nonconformity was given in both cases.

Bear in mind though that *all* results of audit shall be brought to the attention of the relevant staff. Make sure that they hear the good news about compliance as well. Indeed, if an area has had no nonconformity you should make sure that this too is reported back to the staff involved.

A system for verifying the completion of corrective actions is also essential.

Example

A grower and packer of produce had internal audit systems in place but when it came to results of the audits there was no procedure in place to define time frames for corrective actions and verify completion of them.
A nonconformity was given.

The Auditor will also look to see if corrective actions have indeed been carried out in the timescales agreed. Any slippage here might also lead to a nonconformity. You may find it useful to create corrective action forms that include all the necessary features such as nonconformity description, action to be taken, date for completion, person responsible, closure and signature of person responsible and authorisation by the internal auditor.

Example

A relatively large manufacturer of flavourings and additives had a scheduled audit system with reporting of nonconformities. However, a nonconformance raised at an internal audit carried out in February of that year had still not been signed off as completed by the time of their audit (in September of that year). This was considered an unreasonable length of time by the Auditor and an example of poor recording.
A nonconformity was raised.

Part Two

Clause 3.4.4

This is a new clause for Issue 6. Hygiene inspections or audits are an essential part of food manufacturing, and have been for many years. In the previous issues of the Standard, they were not mentioned specifically along with the rest of internal auditing but it is the logical place. The clause refers to the factory environment, including building fabric and equipment. The Auditor will expect to see records of inspections of housekeeping and hygiene and building fabric and equipment condition. Note that inspections must be carried out at least monthly in open product areas, elsewhere the frequency shall be risk based. If the Auditor finds that the inspections are very infrequent in areas where product is contained, you might need to provide evidence of a risk assessment to justify this.

Example

At a manufacturer of fresh produce, the Auditor found that there were no regular audits of the packing building fabric and overhead structures to determine potential risk.
A nonconformity was given.

In other cases, a system might be in place but might not be comprehensive.

Examples

At a manufacturer of cakes and muffins, it was found that building fabric audits did not currently include the filo pastry process area.

At a manufacturer of yoghurt, it was found that the weekly hygiene audit did not include overhead structures (for damage or visual cleaning).

Nonconformities were given in both cases.

The Auditor will, of course, be conducting an audit of the site during the tour and any damage or wear and tear to the structure or equipment that is a contamination hazard will not only be recorded as such but may also tell the Auditor that your auditing system is not effective.

Summary of Internal Audit

Ensure that you have a comprehensive audit system that includes both systems and hygiene and fabric audits. As with HACCP, if you get this right, it will underpin your FSQMS.

Clause 3.5: Supplier and Raw Material Approval and Performance Monitoring

Key points in Clause 3.5
Supplier approval: products and services
Risk assessment
Raw materials and packaging
Audits or questionnaires?
Exceptions: customers and agents
Checking incoming materials
Outsourced production: what does this include?

Clause 3.5.1: Management of suppliers of raw materials and packaging

This SOI concerns your purchasing systems and supplier approval and the objective is to manage the risks involved in raw materials and packaging.

A key word here is 'managed'. The Auditor will seek to establish that purchasing is done in a controlled way, for example that only authorised personnel may purchase goods and services and that there are protocols established for such purchase. The Auditor may ask to speak to someone who is authorised to purchase within your company and check whether they have an understanding of any potential risks with the raw material or packaging.

> ### Example
>
> At a vegetable cannery, the company aimed to control all its purchasing processes, which are critical to product safety, legality and quality, to ensure products and services procured conform to defined requirements. However, the packaging suppliers were not covered.
>   A nonconformity was given.

Clause 3.5.1.1

This is a new clause for Issue 6 and it requires you to carry out a documented risk assessment on *all* your raw materials, which must include consideration of potential risks from allergens, foreign bodies, microbiological or chemical contamination and any other hazards for finished product. You will also need to show a connection between the results of this and your supplier approval and monitoring system. In other words, the greater the potential risk, the greater should be the scrutiny of supplier. The Auditor will expect to see records and might choose several ingredients to study.

Note: Although this clause does not refer to packaging you must include packaging suppliers in the risk assessments.

The concept of risk assessment here should help you to optimise your efforts and focus on suppliers of products where risk is greater. The Auditor may want to see

Part Two

evidence that you are paying closer attention to suppliers of high-risk materials, for example. The risk analysis should lead to the frequency at which the monitoring is to take place.

Importantly, this risk assessment shall also be used to determine the acceptance procedures for the raw materials (see Clause 3.5.2.1).

Clause 3.5.1.2

This clause is derived from two earlier clauses with some new material for Issue 6. Firstly, there is a straightforward requirement that you shall have a supplier approval and continuous monitoring programme. Note that there are certain aspects you should look for, your supplier's traceability being one of them.

The Auditor will be looking for documented procedures that are clear and that are available to all those approved to purchase goods and services. The programme must be comprehensive and include all suppliers.

Examples

An egg packer and processor had a documented supplier approval procedure but the programme set out within it did not include any of the non-food suppliers.

At a vegetable processor, aseptic bag and bottles suppliers did not appear in the approved supplier list.

A packer and grader of herbs, spices and curry powders had a documented supplier approval procedure, based on self-audit questionnaires. However, it did not specify the frequency of sending out the questionnaires (nor what action is to be taken if there was no HACCP in place at a supplier).

A nonconformity was given in each case.

Remember also to include all food contact materials.

Example

A manufacturer of whole and portions of chickens had a supplier approval system but had not included the supplier of dry ice that was used in part of the process.
A nonconformity was given.

You have three options for your supplier monitoring: (1) carrying out audits yourself, (2) relying on third-party audits or (3) using questionnaires.

Whatever method of approval and ongoing assessment that you have set out, there is a need to have a programme and to demonstrate that you are adhering to it.

Examples

A manufacturer of fresh and frozen pasta ready meals had a supplier assessment and review system in place based upon self-audit questionnaires but the Auditor found that it had lapsed due to the lack of response from two suppliers, which had not been followed up.

A bakery used BRC certification as the means for supplier approval; however, they had not confirmed that the supplier had a current certificate.

A nonconformity was given in each case.

Sometimes, a company is part of a group and their suppliers are approved centrally as a group function. It important that companies retain some input at least and some responsibility for supplier monitoring.

Example

A UK-based manufacturer of cocoa-based products was part of a large group with a head office and central technical function in Europe. An approval system for suppliers was in place centrally but it was clear to the Auditor that it was not understood fully by the factory. For example, the audit 'scores' from Europe had not been risk assessed by the UK site.
   A nonconformity was given.

Some companies audit suppliers, others rely on questionnaire-type systems and some use third, party certification. The important thing is to have something in place that is working.

Example

The same manufacturer of frozen Chinese-style products (see aforementioned Examples) had a supplier questionnaire system set up but it was not operational at all; also raw material product specifications were missing.
   Another nonconformity was given.

Finally note that if you go the questionnaire route they must be renewed every 3 years at least and such suppliers must be instructed to inform you of any significant changes.

Question: Will I comply if I have sent out renewal questionnaires but have not received them back within the 3 years?

Answer: No, they must be completed and returned.

Clause 3.5.1.3

This is a clause that quite often trips companies up, so let us be clear what it means. It refers to 'exceptions' and means having a system in place where you allow for the possibility of using products or services where audit or monitoring has not been undertaken by you. This might be because you have to obtain something from an unapproved source in emergency; either that, or where a customer insists on you using a specific source, or finally, if they are purchased through an agent.

Some companies have a strict policy of having no exceptions at all and not allowing any purchase of non-approved suppliers. They may control this by having a number of approved suppliers for each item. If this is the stated policy, the Auditor will look closely at your suppliers to check if there is at least a dual source and be sure that no unapproved suppliers have been used.

The important point is that you have a procedure in place, and as this leads on from the previous clause it is implicit that this too is a documented procedure.

Examples

A manufacturer of chocolate novelty items had a supplier approval and monitoring procedure but it did not define how exceptions are handled when audit or monitoring has not been carried out.

A manufacturer of baked rolls, baguettes and garlic bread had an elementary exceptions procedure but it was not fully documented.

A nonconformity was given in each case.

The Auditor will be looking for a procedure that fully covers this issue. This clause is partly there to allow for the urgent or emergency situation where you need to use an unapproved supplier to ensure continuity. For example, this might be where your usual supplier is unable to supply for a period. If you do purchase from non-approved suppliers in special circumstances, you will need to state exactly what those circumstances are, who is authorised to purchase under those conditions and at what point the normal supplier approval procedure is then invoked. If it is because of customer requirements you must have a system in place that gives assurance of product safety. The Auditor may ask for evidence through your dealings with the customer, their system of approval, for example. If you are purchasing indirectly through agents you will have to show a measure of control, again perhaps using their system of approval.

Clause 3.5.2: Raw material and packaging acceptance and monitoring procedures

The control of the standards of your raw materials and packaging is crucial to food safety and quality, so it is surprising that these subjects have not been covered so specifically in previous issues of the Standard. Thus, this is a new clause for Issue 6 but it clearly concerns issues that are not new. The SOI requires that you have controls in place for the acceptance of raw materials and this also includes packaging.

Clause 3.5.2.1

As stated previously, the risk assessment carried out to meet Clause 3.5.1.1 shall also be used here to determine the procedures for acceptance of raw materials and packaging. You have four options:

(1) Visual inspection
(2) Certificates of conformance
(3) Certificates of analysis
(4) Sampling and testing

You also need to create a list of raw materials with the acceptance criteria (including frequency of testing). Clearly, the Auditor will ask to see the documentation, including your list and that you have actively used the risk assessment to establish the rigour of your testing regime.

Examples

There was no documented guidance, examples (e.g. photographs) for the subjective assessment of in-coming fruit, vegetables (part of CCP1), for example presence of rot, disease, blemish, etc., at a produce processor.

A rice miller was found to have inadequate checks for conformity of packaging and calcium carbonate to specification. Incoming material was signed in but there was no documented procedure of what to check.

Packaging materials were not inspected on receipt at a bakery or recorded as other raw materials as they were currently not part of the standard goods-in process.

A nonconformity was given for each case.

Clause 3.5.2.2

You need to demonstrate that each and every batch of incoming material has been put through your acceptance procedure. Clearly, the Auditor will need to see that you are carrying out any checks that you say you are.

Examples

A manufacturer of frozen Chinese-style products had a documented system that laid down temperature checking of all incoming chilled raw materials but they had failed to meet their own procedures and consistently check temperatures on intake.

At a produce processor, there was no evidence (COA or analyses) to confirm that fruit or vegetable products complied with pesticide residue legislation.

> At a tomato cannery, it was found that the monitoring record for reception of tomatoes was not completed according to procedures.
>
> Nonconformities resulted.

Clause 3.5.3: Management of suppliers of services

This is another new SOI for Issue 6. It concerns service providers, and as a manufacturer using the kinds of services listed in Clause 3.5.3.1 (and surely you use at least some of these) your objective is to ensure that they are appropriate and have evaluated any risks to food safety.

Clause 3.5.3.1

Most companies deal reasonably well with the raw material suppliers. The most frequent problems occur where companies have not included service suppliers in their thinking.

> ## Examples
>
> A manufacturer of mayonnaise and sauces in glass bottles and plastic containers had adequate supplier approval systems for raw materials and packaging but no formal supplier approval was undertaken for their laundry supplier.
>
> A cooked meats factory had adequate supplier approval systems for raw materials and packaging but no formal supplier approval undertaken for the leased final product warehouse facility.
>
> Both the aforementioned examples resulted in nonconformities.

More documented procedures are required here, this time for two aspects: (1) approval and (2) monitoring. Unlike suppliers of actual raw materials, there is no requirement to audit or issue questionnaires but you must demonstrate some form of approval. The Auditor is also going to ask you how you monitor such service providers. This should be relatively straightforward with some service such as cleaning or laundry. You might have to think about how you monitor your laboratory service, for example. Note that the approval method of your laundry service will depend on what production zones you have in place (see Chapter 17, Clauses 7.4.3 and 7.4.4).

Clause 3.5.3.2

You must also have formal contacts for all such services, so expect the Auditor to sample these to have a look at. Two points to bear in mind. The contracts must define what the service is and ensure that any risks to product are addressed in it. A typical example would be the use of non-toxic baits in a pest control contract; equally would be the segregation of clean and dirty clothing in your laundry contract. But suppose your catering service does not have as good in-house allergen control as you do? You might need to address subjects that are specific to your needs.

Clause 3.5.4: Management of outsourced processing

What is this one about? This new clause for Issue 6 concerns the sub-contracting of parts of your process. This is where you may start a process and have to send out to a specialist company an intermediate part. The resultant semi-processed item is then returned to you for finishing. An example is where you might produce a liquid that requires drying before it goes into your final product. Thus, you send it out to a specialist drying company and use the dried material on return. It can include any stage of production so could even include an element of contract packing so long as the product remains under your control and returns to you for final finishing and release. It could, of course, be done by a sister company or a completely independent one. Either way, your objective is to manage the process to ensure overall product safety, legality and quality.

Clause 3.5.4.1

This is very important. You must make it known to the brand owner of any relevant finished product (e.g. a private label owner) if outsourcing is part of your process and you must have their approval for doing so. To demonstrate this, you will need to show the Auditor some documentation.

Clause 3.5.4.2

This clause requires that you must approve the company or companies that you use and this can be done either by auditing them yourselves or ensuring that they have a third-party audit certificate.

Note: Unlike raw material suppliers you may not rely on questionnaires for this one: they must be audited.

Clause 3.5.4.3

There are two aspects of this clause. As usual, you must ensure that you have contracts for the service that detail the processing and the product specification required.

Very importantly, you must also maintain traceability. This will involve a degree of co-operation with your contractor who must ensure continuity of your batch numbers with any systems of their own. If you employ outsourced processing, the Auditor might choose a relevant product to include in their traceability challenge to determine if it is properly covered.

Part Two

Clause 3.5.4.4

What about when the product returns to you? On return, you must have some kind of inspection or testing regime in place. The nature of the testing must be dictated by the risks of the product just as for any raw material entering your site. There may be important quality as well as safety criteria. For example for a dried product such as milk powder, you might need to check moisture content, solubility, colour, presence of black specks and so on.

Clause 3.6: Specifications

This is a very important SOI concerning the range of product and service specifications that you need to have.

Key points in Clause 3.6
Raw materials, packaging, finished products and services
Accuracy and defined limits
Legislation
Matching specifications with work instructions and recipes
Branded product specifications
Specification review

Specifications for raw materials and finished products are generally a part of the culture of most food factories. However, regarding raw materials, the Auditor will be looking at what specifications you hold to see if you have covered *all* raw materials and packaging, and any oversights could lead to nonconformities.

Examples

A manufacturer of jams and blender and packer of honeys did not have comprehensive, documented specifications for the honeys that they were bringing in.

A company specialising in the processing and packing of pasteurised, fermented, baked, cooked and smoked meats had no raw material specifications for beef.

A company was producing a range of potato products and stir-fry meals. The production of the stir-fry products involved the injection of the product with carbon dioxide (CO_2) to freeze it. The company had no specification for the CO_2 to ensure that it was food grade.

In the aforementioned cases, the companies had some raw material specifications but not all, so nonconformities were raised.

The aforementioned examples all concern raw materials but it is also important to have specifications for your packaging and any product contact material (e.g. plastic liners). The Auditor will be looking for examples and later under Clause 5.4.1, the subject of packaging specifications is very relevant, including the requirement that packaging complies with legislation and is suitable for food use. Any specification for packaging should include that information.

Similarly, the Auditor will ask to see samples of certain finished product specifications. They must have sufficient detail to characterise product and confirm that they meet food safety and legal requirements.

Example

At a supplier of Halal meats, the company supplied bulk product for further processing. The current finished product specifications were incomplete and did not include legal requirements or have any authorisation.
A nonconformity was raised.

The other matter is to include any product or service that could affect the integrity of product. The Auditor may ask you about certain services that you use such as contract cleaning, laundry, pest control and so on. You may need to demonstrate that you have specified the service supplied in relevant contracts. See also Clause 3.5.3.2.

Clause 3.6.1

This requirement concerns the adequacy and accuracy of specifications and compliance with legislation. It concerns also the need for defined limits. Certainly, as regards your finished product specifications, it is particularly important that you can demonstrate to the Auditor that you have a good knowledge of the legal requirements of your industry. They will examine specifications for this aspect. They will also look to see how detailed they are and it is important that specifications reflect any changes in legislation. Regarding raw materials and packaging, again, they must be up to date and assure both yourselves and your customers that legal requirements are being met. See also Clause 5.4.1 on packaging.

They will also look to see that specifications contain specific parameters for the necessary chemical, microbiological or physical attributes of the material with clearly defined limits.

Clause 3.6.2

This requirement is about making sure that work instructions and recipes match customer specifications. Thus, any working instructions must include clear details on the product composition, ingredient quantities and any related preparation techniques. If your systems are manual, then recipe sheets will probably be necessary. If you have automated systems including dosing, then you must have instructions in place for the operation of such systems.

Part Two

The Auditor will look to see that you have such work instructions and may sample specifications to see that the recipes used and any related work instructions comply with them.

Clause 3.6.3

Naturally enough, your demanding private label customers will require you to submit and agree very detailed finished product specifications, all in their own special formats. Many of these are Internet-based and you have to do it. As you will know, some of them are very particular and require source information of all raw materials. So it may seem that this clause is like a repeat of part of the SOI, but in fact, as well as ensuring that you have your finished product specifications available, it is also to ensure that if you manufacture branded products also, there are specifications available. It has not always been the case in the past that companies producing a branded product have seen fit to write a finished product specification for it, but it is now a new requirement of the standard to have them and customers do often require them. In this case, customers might be other businesses such as catering institutions, which is why it has become more common practice. While it is not expected that they would necessarily be as detailed as a typical private label specification, they should be at least of a kind that an auditor can verify the products meet legal requirements and that that a customer understands the safe usage of the product.

Clause 3.6.4

This is sometimes a troublesome clause. Where specifications are in hard copy, the Auditor will expect to see some evidence of formal agreement such as a signature or exchange of letters. The clause refers to agreement with 'relevant parties'. The Auditor will want to see formal agreement with your customers on your finished products specifications, especially if they are under private label. On the other hand, if you are selling a branded product but have a specification document that you issue to customers for information only then a formal agreement may not be necessary.

> **Example**
>
> A company that cut, sliced, grated and packed hard cheeses completely failed to meet this requirement: specifications for ingredients and finished products were not signed or formally agreed.
>   A nonconformity was given.

Regarding private label products, for example retailer's own labels, many specifications are now in electronic form, often using proprietary systems that, certainly in the United Kingdom, have been taken on by the major retailers. Many companies will be familiar with these systems that require entry of information through the Internet. They are wonderful tools for the retailers, while a little demanding for the companies who have to complete them.

They usually require the input of a large amount of information and one of the problems that has arisen especially with the amount of information that these customers now have to approve is that while companies have completed specifications as requested by their customers, formal acceptance and approval can take some time. The second sentence of this clause reflects this situation. It says in essence that if you have made all possible efforts to have your specifications agreed but are still waiting for your customers to finish the job, then as long as you can demonstrate this to the Auditor, you will be complying.

Clause 3.6.5

This clause requires the review of specifications.

Review should clearly be carried out when there are known changes to formulation. In addition, the Auditor will be looking for a process that would ensure that all specifications are current as regards legal and food safety requirements and in the case of your finished products that they also continue to meet your customer requirements. Even where you have no changes, there is a requirement for review at least every 3 years. This is a new requirement for Issue 6.

Note: This applies to all specifications.

Examples

During an audit, a cider manufacturer was noted to have some raw material specifications dated between 5 and 7 years old, which were considered to have not been adequately maintained.

In an audit carried out also in 2005 on a manufacturer of deep frozen pizzas, not all specifications were reviewed regularly to ensure current status; notably a packaging specification dated 1999.

In a pasta factory, it was noted that there was no review process at all for any specifications.

Nonconformities were given in all cases.

Clause 3.7: Corrective Action (Fundamental)

This SOI is also a *Fundamental Requirement* and we have discussed the reasons for this in Chapter 4. The key objective here is the objective to use information from any failures in systems to make corrections and prevent recurrence.

Part Two

Key points in Clause 3.7
Nonconformity procedures
Timescales
Authorised personnel
Verification
Root cause analysis

Clause 3.7.1

This particular requirement is all about procedures. There are aspects here from several clauses in previous issues of the Standard all bound up in this one, so this is typical of a merged clause. There are a number of bullet points that your procedures must contain, so let us look at these. Firstly, your procedures must set out how to present clear documentation of the nonconformity or issue and you also need to arrange for assessment of any consequences by a competent and authorised person. It would be sensible to identify the person in the procedures.

Of course, corrective action is what this is all about. You need a procedure in place to identify the corrective action, the timescale for correction and the persons responsible for carrying it out. There also needs to be verification of corrective action. Most companies address this with standard forms.

The Auditor will want to see evidence that you have such procedures (as usual, there is a good reason to have these documented) and if possible evidence of such procedures being put into practice. The reason for this being part of a Fundamental Requirement is clear. It should be a well-established system. The chances are that issues will have arisen in the past to provide such evidence that the Auditor can sample.

Example

A bakery supplying baguettes, rolls and garlic bread had a rudimentary procedure for corrective action, but the Auditor considered that it was not sufficiently formalised, relevant staff had not been trained in it and were not sufficiently aware of it.
    A nonconformity was given.

The Auditor will be examining actual corrective action plans for completeness. Likely sources of interest for the Auditor are internal audit, pest control, laboratory test records, production record sheets and so on. It is in the area of poor recording that many nonconformities arise regarding corrective action.

Example

A manufacturer of chilled and frozen poultry products was found to have no record of corrective actions and agreed responsibilities and no timescale documented for issues found by an internal audit carried out some months before.
A nonconformity was given.

The inference in both aforementioned cases is that there had been some corrective action but it had not been documented.

Regarding closure and verification of corrective action, from the Auditor's point of view, it is only possible to be certain that this clause is satisfied if you can demonstrate that you have done so. Ironically, this does mean that the best situation for the Auditor is where you have had some nonconformities and can thus provide hard evidence that you have carried out corrective action. Therefore, a good track record is essential here. If there is no other material for the Auditor, then internal audit reports should be fertile ground for them.

Example

A manufacturer of cakes and muesli bars routinely sent product for microbiological analysis. The Auditor noted that there was no investigation undertaken when an end-of-life microbiological analysis of a pastry product returned a high TVC.
A nonconformity was given.

The Auditor will examine corrective action plans to see that they are sufficiently detailed and that the action to be taken is clear and usually with a deadline. Completed corrective actions should be verified, so the Auditor will look to see that completed ones have been signed off by the person taking action and authorised. In the case of internal audit this should be by the auditor.

Finally, we must say something about our new friend: root cause. You must always identify root cause of a nonconformity and any necessary corrective action (to the root cause), which should result in preventive action.

Clause 3.8: Control of Nonconforming Product

This SOI is a straightforward objective to ensure that all out of specification product is managed to prevent release. Naturally, you may not have such material on site at the time of the audit but the Auditor will need to see evidence that this objective would be

Part Two

met in such circumstances. Examples of quarantine or 'hold' labels would be useful. A defined quarantine area is also a good idea.

Key points in Clause 3.8
Documented procedures
Identify and isolate
Informing brand owners
Decision making
Records

Clause 3.8.1

The requirement is for procedures (documented, of course) to cover the management of nonconforming product. There is a list of seven bullet points to satisfy here and this clause is an example of the merging of three earlier clauses, with some new material added for Issue 6. The Auditor will need to see your procedure, which must include the seven aspects. Clearly, it will be necessary to provide evidence that you are meeting the procedures also.

The first part is a requirement for the procedures to state how staff must react to nonconforming product in terms of identifying and reporting it, so this should be straightforward. Next, there must be clear identification of nonconforming product. This will depend on your systems but can be done by using simple labelling, for example 'HOLD' labels, or you may have a computerised system that enables the blocking of such products. Again, the Auditor will need to see your procedure and will look to see it in action. This may depend on whether you have any material on hold at the time.

Next, you must have some way of securely isolating nonconforming products such as an isolation area, which the Auditor will be able to see, even if not being currently used.

Examples

A manufacturer of soft drinks and coffee products had some out of date coffee on site. The Auditor found that this had not been quarantined, as required in their procedure.

At a produce packer, some nonconforming packaging was being held, unlabelled on a pallet that also held conforming packaging.

Nonconforming product identified during packaging operation at a tomato cannery was not identified in the warehouse and not documented.

During an audit of a canner of pulses, a basket of cans were noted with evidently faulty seams ('peaking' showing). This basket was in the despatch area with no indication that it was in quarantine.

A nonconformity was given in each case.

Next, you must also have a procedure to refer to the owner of the brand if necessary.

Often, a decision will be necessary as to the disposal of the product, for example destruction, reworking or downgrading or acceptance by concession. The procedures must include instructions on these aspects and the responsibility for making the decision must be clearly stated in them. The Auditor is likely to interview staff involved. As always with procedures, there could also be a training aspect for the Auditor to consider.

Examples

At a breadcrumb manufacturer, the Auditor found that some ingredients had gone past their durability date. The company had allowed an extension of date by concession; however, there was no evidence to justify this such as a document from their supplier.

A sauce manufacturer was found to have no 'concessions' procedure documented within the nonconforming product procedure.

At a poultry processor, it was found that where microbiological results showed product to be nonconforming corrective actions assigning responsibility and time frame were not documented, for example positive results for *Listeria*.

A nonconformity was given in each case.

The final two points both concern records; firstly, on the decision on the fate of the product and secondly, records of any destruction of product for food safety reasons. You need a procedure for this and some defined recording system that the Auditor can examine.

Clause 3.9: Traceability (Fundamental)

We come to a very important subject: traceability. Traceability has always been an important issue and has been covered in the Global Standard since the first version. It has had increased focus on it since the publication of EU legislation on the subject (EC 178/2002). The result is that this clause was made a *Fundamental Requirement* when Issue 4 was published. To underline this further, the BRC have a guideline on traceability in their series. The guideline is a good introduction to the reasons for having traceability, to testing your system and gives case studies of both forwards and backwards traceability. The guideline also rightly points out that a traceability system need not be complex or over-elaborate.

Key points in Clause 3.9
Food items and packaging
Secondary packaging: why is this important?
Identification of batches

Testing traceability
Quantity/mass balance
4 hours
Rework

It is worth looking at the precise wording of this SOI: 'The company shall be able to trace all raw material product lots (including packaging) from their supplier through all stages of processing and despatch to their customer and vice versa'. The SOI refers to product lots. To meet this requirement, you will need to have internal batch traceability. In doing so, you should also have both forward and backward traceability. Note that the reference to packaging refers to all packaging that can have an impact on food safety. So, obviously, this includes primary packaging and product liners. It might also include secondary packaging where this can affect the product. In 2011, researchers in Switzerland found that mineral oils in printing ink from recycled newspapers used in cardboard cartons can migrate into food, even through the plastic inner liners. This led to several UK breakfast cereal companies making changes to their packaging.

Despite the concept of traceability having been with us for many years, it continues to be a source of nonconformity. Many have systems in place but pick up nonconformities because their system was not comprehensive. For example, it appears to be easy to overlook something such as primary packaging.

Examples

A manufacturer of poultry products had a traceability system, but it did not currently include primary packaging or dry ice. The company picked up a further nonconformity because the Auditor considered that there was inadequate traceability labelling of chilled chickens in metal trolleys.

A cake manufacturer had a traceability system that did not include traceability for bulk chocolate.

At a manufacturer of potato-based products, it was noted that the blue polythene liners within the bulk cardboard boxes were not currently being traced.

At a fruit juice manufacturer, the Auditor noted that not all processes had traceability in place for primary packaging, in this case juice bottles and caps.

At a preserves manufacturer, the Auditor found that traceability was not implemented for: glass jar lids, portion pot sealing foil, squeezy plastic bottle caps, vinegar, and several ingredients controlled by the laboratory.

At a manufacturer of cooked meats, it was noted that the gases used for the modified atmosphere packs were not currently traced.

A manufacturer of meat products had performed a recent test but had not included primary packaging.

A manufacturer of produce failed to include both primary packaging and Dry Wite (preservative) in their traceable system.

At a manufacturer of sandwiches, the Auditor found that the traceability system did not include either film or flow-wrap packaging.

All the aforementioned cases resulted in nonconformities against this clause because although the companies did have traceability systems they were not complete.

The examples demonstrate how comprehensive your traceability system needs to be. Most factories have a system of some kind in place but it is easy to overlook something. The Auditor will observe traceability in action, particularly during the factory tour when examining on-line records. The effectiveness of your procedures will depend on how well your staff apply them.

Clause 3.9.1

This clause requires that all materials that need to be traced are adequately identified. This may take the form of the simplest system of batch numbering. It is vital that your system is comprehensive, formalised and properly implemented.

Examples

At a bakery, primary packaging materials and Dry Wite were not currently traced (i.e. no trace batch/lot code records). Malthouse and vitamin C are not formally included in the bakery traceability log (handwritten trace record on reverse of record sheet). Current trace records for Dry Wite had not recorded the correct batch code (but had recorded the previous batch code by mistake).

At a tomato paste manufacturer, traceability of the salt used for product was unclear because the bill contained three different lots of salt.

A nonconformity was given in both cases.

Date codes on raw materials may be useful but the same date code may cover more than one batch. Similarly, you may receive several deliveries from a supplier, on different days and they may all be from the same batch. The Auditor will look to see that your system is as precise as it can be to ensure traceability, so do not necessarily rely on your supplier's systems if you believe that you can refine this further.

Remember to include packaging materials, in particular your part-used material.

Part Two

Examples

At a bakery, there was no adequate system to assure packing traceability.

At a vegetable canner, part used packaging was covered before being returned to storage, however, it was not labelled and thus traceability was lost.

A nonconformity was given in each case.

Remember also that your intermediate products, such as sauces that may be used in several batches of finished products, need to be suitably identified.

Example

A raw fish processor had a traceability system that relied on colour-coded stickers being put on to polystyrene cases as the fish were received into the chill store. Each colour represented a different day of the week. During the audit, it was found that two different colours were being used that day and some cases had not been labelled.
A nonconformity was given.

Any lot numbering system will only be worthwhile if the personnel use the system in their records.

Example

At a vegetable canner, traceability was effected by (1) lot number of processing aids, which is not recorded in the form and (2) lot number of semi-finished products, for which, the operators were observed to forget to write down.
A nonconformity was given.

Clause 3.9.2

This should be a relatively straightforward requirement in that you must test your traceability system from raw material to finished product and vice versa. In other words, you must demonstrate both forwards and backwards traceability.

This must also include a check on mass balance quantities to ensure that all raw material has been accounted for and reconciled.

The requirement is that the tests are done at a predetermined frequency and that this shall be at least annually. Finally, you should be able to achieve full traceability within 4 hours.

The Auditor will want to see a fully documented test result as evidence with any anomalies or nonconformities and consequent corrective actions also documented.

The Auditor also needs to make a judgment on whether you can achieve traceability in 4 hours. They may not be able to do this during their traceability challenge because of the simultaneous vertical audit. Therefore, it might be a good idea to document this in your test records.

Clearly, if you fail to do any testing, a major nonconformity will be given.

Example

A manufacturer of jams and preserves was found to have no traceability test performed or documented from final product towards raw materials, or from a raw material towards final products.

Thus, a major nonconformity was given.

Nonconformities can also be picked up if your test is incomplete or if you miss a deadline.

Examples

A manufacturer of prepared produce had performed no in-house traceability tests during the 12 months prior to the audit.

At a prawn supplier, there was no documented backward traceability test from a final product towards all ingredients and primary packaging in the previous 12 months.

A nonconformity was given in each case.

Furthermore, the Auditor must carry out a test during the audit. At this point, they will be looking for both pure traceability and will also use the exercise to carry out a vertical audit of your systems (See Chapter 10).

Considering the traceability test itself first, they might bring a product along to the audit that they have purchased, or they might select one from your storage areas or retained samples. They will almost certainly choose one that will challenge your systems fully, for example one that has the most ingredients. They are unlikely to choose one much older than 5 months.

As the examples mentioned previously show, your system must include all ingredients, bulk materials and, as far as possible, other materials that come into contact with product, for example gases and, of course, the packaging. The Auditor will be aware of all your ingredients and will be looking for traceability of actual batches of each item.

In addition, the Auditor will also ask for relevant raw material specifications and all associated process control records for production of the finished product, thus comprising a vertical audit. The type of records that you are likely to be asked for will relate to raw material intake, recipe control sheets, process control records, packaging records and dispatch records as a minimum.

Example

At a manufacturer of bacon, the Auditor carried out a traceability test during the audit. The D cut rindless gammon could not be traced back to the raw material.
   The Auditor gave a nonconformity.

The Auditor will also do a forwards test and pick examples of raw materials and see if you can account for which products they have gone into and where those products were dispatched to. They will also ask you to confirm quantity check/mass balance.

Question: Do I have to do quantity check/ mass balance on all my raw materials?

Answer: No, you need to sample every year.

Clause 3.9.3

Rework is an important subject and is sometimes overlooked in traceability systems. For some companies, rework appears to be complicated in this regard, but there should always be ways of relating rework material to its components. Where rework is continuously fed back into systems, then traceability may be already assured. Where rework is segregated in batches, then it becomes an 'ingredient' with its own traceability history. Surprisingly, this does sometimes cause problems, either where companies fail to consider it or where their systems are not complete.

Examples

A jam and honey manufacturer had a sister company making similar products. Material for rework was sometimes taken in from the sister company but it was not always traced.

In a cider manufacturer, a rework system was in place and traceability had been considered, however, the reworked product was not currently fully traceable as work order numbers for the source of the rework were not recorded.

At a vegetable canner, it was observed that the storage of semi-finished product in production halls and reworked product of the same lot were not labelled.

A nonconformity was given in each case.

The Auditor will want to see that batches of rework have their own batch control systems.

Finally, regarding rework, you must also ensure that product legality and safety is not affected by the practice. For example, you must consider that any ingredients declaration remains accurate if ingredients are re-worked into a product. Similarly, that your allergen control or any identity preserved products are considered. The Auditor will be conscious of this and will look closely at any re-work activities both during the document assessment and during the site audit (See also Chapter 15, Clause 5.2.5 relating to allergens).

Clause 3.10: Complaint Handling

The company that has no complaints is a rarity, although some claim to exist. Nevertheless, all companies must have a system for the management of complaints. The BRC has a guideline on Customer Complaint Handling, which contains some good advice on what your system should include as a minimum and gives guidance on how to respond, investigate and take preventative action.

A complaint system should be integral to your FSQMS and the Auditor will expect to see it referenced in your Food Safety and Quality Manual and to see formal procedures. The simple objective in this SOI is the reduction of recurring complaint levels.

> **Key points in Clause 3.10**
> Investigating complaints
> Root cause
> Response times
> Records
> Training
> Trends

Clause 3.10.1

You must record and investigate all complaints. The Auditor will ask to see your actual complaint records and will assess both the seriousness of your complaints and how you have handled them. At the simplest level, the Auditor will expect to see at least a complaint log.

Examples

At a small ice cream manufacturer, the Auditor saw evidence of corrective actions from complaints but the original complaints themselves had not been recorded.

At a Danish wine bottling plant, the Auditor found that while complaints were being dealt with, there was no formalised complaints procedure.

A nonconformity was given in each case.

Part Two

Your records need to be complete and show that they are all logged and give detail of any relevant investigations.

> ### Example
>
> At a bakery, there were no documented customer complaint investigation reports.
>   A nonconformity was given.

Furthermore, you must also investigate the root cause of the issue (see Chapter 19 for more on root cause analysis). Note that this is only required where you have sufficient information. This is recognition of the fact that good information is not always forthcoming from either consumers or retail customers!

Your action must be prompt and efficient. There is also a training element here in that your actions must be carried out promptly and by trained staff. The Auditor will look at actual complaints received and ask how quickly you were able to respond. Correspondence and replies back to customers should be retained so that this can be assessed. Furthermore, corrective actions taken by you must also be carried out efficiently. The Auditor will be looking to see that your response to complaints is governed by the seriousness of them. They will ask who is responsible for complaint action and enquire about their training.

Note that the clause also requires that staff are appropriately trained to carry out the necessary action.

Clause 3.10.2

No one welcomes complaints but they should be used as a good source of information for improving your systems. The Auditor will also be looking at them with this in mind. For example, if a high number of foreign body complaints is noted, the Auditor may use this to question whether your prevention of contamination is sufficient, or they may review your hazard analysis to see if this aspect is assessed sufficiently. It is better if you have done this first!

The clause specifically concerns general, on-going improvements rather than specific corrective actions to individual complaints. The Auditor will ask to see statistical or trend analysis of complaint data, which is a useful tool. They will ask how you use this data and whether you present it to staff in some way, for example. The requirement is that the analysis is made available to relevant staff.

> ### Example
>
> At a rice flour manufacturer, the Auditor noted that complaints were not fully investigated or communicated effectively across different departments.
>   A nonconformity was given.

Remember that your overall objective is to reduce recurring complaints. One question that sometimes arises is whether you need to include complaints in your statistics if they have been caused by customer 'abuse', for example by not following instructions or by using the product after its durability date. The answer to this is that if such complaints are repeated, a root cause analysis should be undertaken to see if there is an issue with product labelling that may lead to this situation.

Clause 3.11: Management of Incidents, Product Withdrawal and Product Recall

There can't be a crisis next week. My schedule is already full.   (Henry Kissinger, *New York Times*, 1973)

This is another very important SOI and, like traceability, is part of EU legal requirement (EC 178/2002). Your objective is to be able to manage incidents by way of predetermined planning and to have effective product withdrawal and recall procedures in place.

Key points in Clause 3.11
Contingency planning
Business continuity
Documented procedures
Having all contacts details
Testing the system
Informing the CB

It is important to be clear about the terms used here. Definitions for these terms are given in the Glossary to the Global Standard.

'Withdrawal' should be taken as being the removal of relevant product from sale and managing all existing stock of an item so that it is isolated and brought back under the control of the company. In the case of retailer customers, this would mean taking all stock out of shops, storage and distribution systems.

'Recall' includes all the aforementioned, but, in addition, the retrieval of affected product already sold through to the end consumer. This may require public notices etc.

The BRC have produced a guideline to product recall that is a stand-alone document (not part of the Global Standard series). It is a very comprehensive document and well worth having.

Clause 3.11.1

This requirement is to have procedures for the management of incidents and to have effective withdrawal/recall procedures. The Auditor will be looking for documented procedures here and would expect them to be also referenced in your Quality Manual. They should be detailed enough so that in the event of an incident any key member(s) of staff would be able to carry out a withdrawal or recall efficiently.

Part Two

> **Example**
>
> At a manufacturer of sausages, haggis and beefburgers, the Auditor considered that the current documented procedure on product recall did not ensure adequate reaction by key staff in the absence of the director or their consultant.
>   Clearly, in the event of a crisis, the system should cover any absence of key staff.
>   A nonconformity was given.

This requirement is not only about isolated incidents that might lead to product withdrawals, but also general crises. Thus, your procedures might also include contingency plans in the event of forces majeures such that your business continuity faces interruption.

Note the requirements in the clause for what kind of situations you must include in your procedures (a change from Issue 5 in which they were for guidance only). The first of these, the disruption to key services is probably the most familiar to us. For example, you should consider what happens in the event of an electricity power cut. This would be especially significant if your systems and documents were all computer based.

If you have had any incidents, the Auditor will ask to see records of how you dealt with them so that a judgment can be made on how well you managed them. Your records of any incidents will, therefore, need to be clear and detailed and include corrective and preventative actions.

Clause 3.11.2

This clause comprises elements of several clauses from Issue 5 of the Standard. It is about the specific procedures you shall have in place for product recall.

You must have:

- identification of key personnel constituting the incident management team;
- guidelines for decision making;
- a list of up-to-date relevant contacts and a communication plan including customers and relevant authorities plus support agencies – note that this must also include your CB;
- a communication plan so that customers and all relevant persons are notified in a timely manner (for 'timely' read 'quickly'!);
- details of external agencies, laboratories and other sources of expertise; and
- a plan for the logistics of it all.

It is important that this procedure is acted upon. The Auditor may judge your compliance with this by asking to see records of an actual incident and examine how quickly you did inform customers.

The incident management procedure must include the withdrawal or recall procedures. It is important to convince the Auditor that you will be able to account for all affected products. When examining records of an incident or a test, the Auditor will ask to see how you ensured that you knew the quantities involved and were able to account for all.

The procedures should also show your awareness of your obligations to the public and to the authorities. For example, under EU law (EC 178/2002), you must inform the authorities of any food that has been placed on the market that is injurious to human health.

Example

A UK-based manufacturer of frozen pies, sausage rolls and pasties for cooking had a product withdrawal and recall procedure but the product recall procedure did not include the requirement to advise the FSA in the United Kingdom or their CB of a recall.
A nonconformity was given.

It is also very important that your plans are kept up to date. Thus, the Auditor will be conscious of any recent key staff changes and be looking at your list of key contacts both internal and external to see if it is current.

Examples

A liquid milk dairy had a documented product withdrawal and recall system. However, they had undergone some significant re-organisation of their management structure some months before their audit. It was noted that the procedure had not been revised since the organisational changes.

A nonconformity was given.

At a manufacturer of modified maize flour, it was noted that timescales for notifying customers of potential recalls in a timely manner were not defined. While there was no evidence that customers would not be informed quickly, the Auditor felt that the procedures required more detail to ensure this.

Thus, a nonconformity was given.

Clause 3.11.3

The concept of product withdrawal and recall was brought into sharp focus in the United Kingdom in 2005 when an incident involving Sudan 1 red, an artificial colour that is not approved for food use, was found in a wide range of food products. This had come about because it had been used to colour chilli powder that had been used in

Worcester sauce, which in turn had been used in many recipe products. This incident truly tested the traceability and recall systems of many manufacturers and retailers and in total resulted in the largest product recall in UK history. The BRC were very closely involved in assisting that incident and hence their guideline to product recall is very well informed.

Therefore, it is considered important that you verify your own withdrawal and recall systems by regularly testing them, hence this requirement.

The Auditor will be firstly looking to see if you have scheduled and carried out a test at all. I think of this requirement as one of the trapdoors in the Standard because to fail to carry out a test, and many do, will result in a major nonconformity. It appears to be an easy requirement to overlook.

Examples

An iceberg lettuce manufacturer gave cause for concern for the Auditor in that they had no system for regularly testing their procedure.

At a bakery, there had been no documented annual test of the product recall procedure for over 18 months.

A nonconformity was given in each case.

This clause also creates problems for some companies in that clearly there is no way to carry out a test that is completely realistic without doing a real withdrawal or recall. What you need to do is create a simulation that is as close as possible to the real thing. The BRC guideline gives some very good advice on how to go about this. In some cases, a 'paper' exercise might include forward traceability. Of course, traceability is an important part of this procedure and you must be able to trace all 'affected' product dispatched.

Example

At a manufacturer of frozen Chinese meals, a test of the recall system had been recently carried out; however, the test failed to identify all dispatched product.
A nonconformity was given.

Indeed, many companies carry out their traceability test and withdrawal/recall test simultaneously as it is efficient to do so. However, even if the forward traceability is sound, this may not be sufficient if the Auditor is not satisfied that other aspects of withdrawal/recall, such as alerting customers, are not covered.

Part Two

Example

A produce manufacturer had carried out a supposed test of its withdrawal/recall systems using their traceability test only. The Auditor considered that this did not satisfy all requirements of withdrawal/recall because it did not cover the recovery of product from customers.

 A nonconformity was given.

Assuming you have indeed tested the system, for the Auditor you will need to have evidence available of the test. This might include records of memos to staff describing the mock situation, minutes of meetings, 'advice' to customers, the tracing of products affected to all possible customers and so on. Note that this is one clause that specifically requires that you retain the results of the test. Also, you must specifically record the timings of all the key activities.

Examples

An ice cream manufacturer had carried out a withdrawal/recall test but the Auditor considered that the recall exercise was not fully documented.

A tomato canner has a recall test system but the recall test did not document the timing of key activities.

A nonconformity was given in each case.

Finally, just as for traceability, make sure that you test the system at least annually.

 In the case of a test or the result of an actual withdrawal or recall that reveals problems with the system, the Auditor will look to see what corrective action you have taken or improvements made to preclude errors and ensure that your system can be applied successfully. It is a requirement that following a test or of an actual recall (not withdrawal) the results should prompt a review of procedures.

 The following examples resulted in nonconformities against this part of the clause.

Examples

At a pasta meals manufacturer, there was no formal review after a product recall test.

At a manufacturer of maize flour, an informal review was conducted following a product withdrawal/recall test, but the Auditor considered that this needed formalising and documenting.

Clause 3.11.4

For any genuine recall, in addition to informing the appropriate authorities, you must inform your current CB. A new feature for Issue 6 is that this must take place within 3 working days of the incident. The Auditor will look to see that you have this built into your procedures. Thus, as required in Clause 3.11.2, you must have your CB listed as a contact.

The Auditor may well be aware of any recalls that you have had, so if you did not inform your CB at the time of any recalls this could be a nonconformity waiting to happen.

Summary

The predominantly green coding indicates that this section is largely about document review for the Auditor. It is all about systems and management. But there is plenty for the Auditor to observe in the factory as well that may have a bearing on clauses in the section. Documents being used in the factory, such as work instructions and record sheets, will be examined and traceability identification will be noted.

Part Two

Quiz No. 8

(1) Which clauses mention root cause?
(2) How long shall records be kept in addition to the product shelf life?
(3) How often should you inspect for hygiene and fabrication in open product areas?
(4) How often must you re-issue supplier approval questionnaires?
(5) Which are the three Fundamental clauses in Section 3?

14   Clause 4: Site Standards

Clause 4 of the Global Standard is somewhat different from the previous sections for the Auditor because so much of it is involved with the physical attributes of the site. Much of it will be dealt with during the site tour; thus, you will see in Issue 6 that much of it is colour coded as peach. Some of the clauses in this section should not need much explanation: it should be clear what the Auditor will expect to see. The Auditor is trained to spend a good amount of time in the production areas, 50% of the audit in fact and will want to see all areas and equipment that relate to the scope of the audit. This will include a tour of the perimeter, the raw materials intake, processing areas, packing areas, storage and transport. This section contains two Fundamental Clauses: 4.3 and 4.11.

Part Two

Production Risk Zones

I want to start by referring you back to the concept of the Production Zones that we looked at in Chapter 6 (see also Appendix 2 of the Standard) and consider how they affect your thinking and your compliance. The new way of looking at this for Issue 6 might mean that you have had to redefine certain areas and that you now have new requirements to consider. For example, when considering protective clothing (Clause 7.4), you will need to consider that you might have areas of different risk and your clothing will relate to those specified areas, for example low risk or high care. Under Clause 7.4.3, it might be that you have enclosed product or low-risk areas and protective clothing in those areas is needed only to protect the employee from the products rather than the other way round. If that is the case, it could be permissible to allow them to launder their own protective clothing.

All the clauses refer to all production zones unless they specifically refer to one. There is one clause in this section that refers exclusively to low-risk areas, Clause 4.3.4: Concerning process flow. The following Table 14.1 shows the clauses that specifically refer to high care or high risk or both.

The BRC Global Standard for Food Safety: A Guide to a Successful Audit, Second Edition. Ron Kill
© 2012 John Wiley & Sons, Ltd. Published 2012 by John Wiley & Sons, Ltd.

Table 14.1   Clauses that refer only to high-care or high-risk areas.

Clause	High care	High risk	Subject
4.3.5	✓		Physical segregation
4.3.6		✓	Physical segregation
4.4.4	✓	✓	Drains
4.4.13		✓	Changes of filtered air
4.8.4	✓		Entrance and changing facilities
4.8.5		✓	Entrance and changing facilities
4.11.5	✓	✓	Dedicated cleaning equipment
7.4.4	✓	✓	Laundry of protective clothing

Note: In the clauses, the words 'high care' and 'high risk' are always emboldened.

Clause 4.1: External Standards

This SOI concerns the location and maintenance of your site. Frankly, if your location is poor and does not enable the production of safe and legal products, there is little that could be done about it in the timescale of a Global Standard Audit.

The Auditor may want to see a plan of the site, indeed most CBs ask for site plans to be sent in before the audit takes place. Naturally, the Auditor will be walking the site both inside and outside and will assess the overall suitability of the location.

Key points in Clause 4.1
Looking after the perimeter
Local activities: are they a problem?
Your road surface
Birds roosting
Building fabric

Clause 4.1.1

This requirement concerns any local activities that may affect your products. For example, there may be something that your neighbour does that could impact on your site. The Auditor will ask what type of activities are going on in your vicinity and will also observe this during the site tour. There are many potential problems from neighbouring activities. For example, areas of concern could be the nature of neighbour's products, waste, dust, smells or effluent. It could be that your neighbours have scant regard for potential pest problems for example. Any of these that could spill over on to your site might be an issue.

If there is a problem relating to any external factor such as flooding and there is a potential risk to your product, then you will have to consider measures to prevent contamination. In such circumstances, the clause requires also that you review such measures regularly and the Auditor would need to see evidence of this review.

Clause 4.1.2

This clause is about maintenance of your exterior and calls for general tidiness and maintenance. It has been merged from two previous clauses. It also concerns in particular the maintenance of grassy or planted areas. This is of importance because of the potential for the harbourage of pests in unkempt areas. The Auditor will walk around your perimeter and look at all the areas that are accessible. The Auditor should take a reasonably practical view of this; your site is not a showpiece after all but a working environment. However, a poorly maintained area not only may be a danger to product or personnel but it also shows a lack of attention to your immediate environment that can have many consequences. Perhaps in the same way that an untidy desk shows an untidy mind, a lack of attention to the general conditions outside may indicate to the Auditor that all is not well on the inside either. If any areas appear to be overgrown or neglected, you will be given a nonconformity.

Examples

At a fresh sausage maker, the rear perimeter wall and passage were obstructed by waste materials.

A canner of tomatoes, pulses and sauces was found to have wooden pallets stored near the entrance to the finished product warehouse.

At a bottler of natural mineral water, it was noted that there was an unacceptable build-up of debris (machine parts and old pallets) on the back wall of the yard.

Outside a bakery, disused engineering items were stored adjacent to the 'green' container in a manner likely facilitate pest harbourage.

A manufacturer who specialised in washing and packing potatoes was found to have their perimeter wall at the rear of the building obstructed by obsolete equipment.

At a produce processor, it was found that the area between the new compressor room and the new extension was overgrown and cluttered with rubbish. There was also surplus lagging stored on the east wall.

At a manufacturer of herbs, spices and curry powders, it was noted that grass to the rear of building was overgrown against the wall.

A nonconformity was raised in each case.

In all the aforementioned cases, the Auditor would have considered that the conditions of the perimeters posed a potential threat of pest harbourage and did not lend themselves to ease of pest control or general maintenance of the site.

Finally, your traffic routes must be well surfaced and be kept in good repair. So, no potholes collecting water, please.

> **Example**
>
> At a manufacturer of fresh sausages, the Auditor found that the forecourt was in a poor state of repair and was retaining surface water.
>   A nonconformity was given.

Such exterior maintenance should be scheduled into your procedures.

Clause 4.1.3

This is a requirement to ensure that the fabric of your buildings is in good repair and in particular that items such as pipe work are properly sealed. It is a good idea to include this aspect in your internal audit.

> **Examples**
>
> There was a 1-cm hole in the production area wall at the end of the glass bottling line, leading to the exterior.
>
> There was a disused drainpipe on the wall of the warehouse, leading to the exterior and allowing possible access for pests.
>
> A nonconformity was given in both cases.

If you have intake pipes, for example for liquids or powders, make sure that they are sealed when not in use. For certain intakes, it is a good idea to have locking caps as a preventive measure and an Auditor may pick you up on this if they are not so.

There are several possible examples here that the Auditor is likely to look for. Any external storage in bulk silos should be well sealed and protected. Intake points should be closed off when not in use and preferably locked to prevent any possibility of tampering.

> **Example**
>
> A medium-sized bakery had external flour silos. They were in good condition but the Auditor found that the intake points were not adequately protected or lockable.
>   A nonconformity was given.

A similar situation may apply even where the actual storage unit is inside, if the intake is external.

Example

A medium-sized bakery in London had external flour silos. They were in good condition but the Auditor found that the intake points were not adequately protected or lockable.
   A nonconformity was given.

Another similar situation sometimes arises with dairies where flexible pipework is used to hook up to the milk tankers and where they are not suitably protected against possible contamination. They may be left on the ground, sometimes open at both ends.

Example

A milk dairy had uncapped and unprotected raw milk intake tanker connection hoses, externally stored when not in use.
   A nonconformity was given.

When I started in the food industry, too many years ago to mention, the roosting of birds outside factories in the United Kingdom was a significant problem. It was not unusual to arrive at a bakery or flour mill and see rows of pigeons and sparrows lined up on the roof. In the United Kingdom, it has certainly reduced greatly as a problem compared to those days, but wherever you are, you must also ensure that you prevent the potential roosting and nesting of birds.

Clause 4.2: Security

Security continues to be an area of concern. There have been several incidents in recent times of deliberate contamination of products. These can arise out of simple mischief making, disgruntled employees, blackmail attempts and so on. In the recent past, there was also a real fear, certainly in the United Kingdom, of the possibility of terrorist activity affecting food supplies by this kind of sabotage.

The SOI requires that security systems protect the product from malicious contamination. The subsequent clauses enlarge on this; they were significantly revised for Issue 6.

Key points in Clause 4.2
Preventing malicious damage
Annual review

Part Two

Staff training
Defined areas
Visitor reporting systems
Registering your company

Clause 4.2.1

To meet this requirement, you must have carried out an assessment of your security arrangements. In particular, you must consider the potential for deliberate contamination or damage. The Auditor will want to see documents to confirm this and they will also have this in mind when conducting the site tour both internally and externally. Note that the clause also requires that you define sensitive or restricted areas and that they are 'clearly marked monitored and controlled'. This last phrase suggests that such areas will need signage to indicate them, but there may be other ways to meet this aspect such as clearly drawn up plans; however, remember the purpose of this requirement which is to prevent unwanted visitors or unauthorised staff gaining access. Remember to consider any items stored outside.

Note: There is also a requirement to review your measures at least annually.

Clause 4.2.2

The idea here is to maintain security and prevent access to unauthorised persons. The Auditor will hopefully have passed through some security measures on arrival because you must have a visitor reporting system. Note that these signing in systems should be used to encompass all visitors including contractors and your records may be studied by the Auditor to check this. They will have noted how easy or difficult it was to access the site themselves and be aware of your signing-in procedures and so on. They may also ask what special measures you have in place to maintain security at all times. Do you have constant manning, CCTV, secure boundaries and so on? Consider all your access points including staff entrances.

> ### Example
>
> A yoghurt manufacturer had a modern, brick-built factory with a main entrance and a single door as a staff entrance. The Auditor noted that this staff entrance was left unlocked during the day to allow for staff coming and going. However, there were no special controls on this access point and the Auditor felt uneasy about the lack of security here.
>   A nonconformity was given.

You should also be aware of this aspect if you are likely to have doorways open for prolonged periods. This might be the case for an intake door for example. If it is kept open because of time taken unloading you should be aware of the security issues here.

Sometimes, factories also allow staff to open fire doors in production areas to allow ventilation on hot days. This might be acceptable from the pest point of view if there are sufficient pest barriers in place but be aware of the potential security issues.

There is also a requirement to train staff in such security measures and this is essential. A key point here is the requirement to encourage staff to report unknown visitors. Remember that the Auditor, a visitor, will be in an ideal position to test this one!

Note: The requirement in Issue 5 to challenge unknown visitors has been removed.

Clause 4.2.3

Many countries, including the United Kingdom, require that food premises be registered with the local authority. Where this is the case, be sure that you have registered and have the evidence available for the Auditor.

Clause 4.3: Layout, Product Flow and Segregation (Fundamental)

This SOI is a *Fundamental Requirement* because it concerns the basic layout and maintenance of the premises and plant. A major nonconformity against this could not realistically be resolved within a reasonable timescale, hence the fundamental status of this clause. Your objective is to safeguard the product through good factory design and the movement of personnel.

Key points in Clause 4.3
The importance of layout for flow of product and staff
Having a good site plan
Segregation: high care and high risk
Visitors and contractors: making them aware
Temporary structures

The Auditor will examine a site plan (see Clause 4.3.1) in order to look at product flow and overall layout, naturally taking in the different production zones and the movement of staff, with a view to seeing how you prevent product contamination by design. This might be of special importance in high-risk production.

The Auditor will also be assessing whether you meet this SOI during the factory tour. During the site tour, the Auditor will be looking to see how you handle raw materials, packaging, intermediate products and finished products and whether you expose product to risk by not ensuring that product and packing is secure. The Auditor will also look at the factory structure and equipment to see whether this facilitates the maintenance of product integrity.

Significant flaws in factory design could lead to a major or critical nonconformity and hence result in a 'No Grade'.

Part Two

Clause 4.3.1

This is a new clause for Issue 6 and is fairly self-explanatory. You must provide a site plan with all your product risk zones identified (see Chapter 5 and Appendix 2 of the Standard if you have not yet read either). Note two further points. You must take into account your product zones when determining your prerequisite programmes. Also, Clause 4.3.2 concerns a site plan and you may wish to have one, which meets both requirements. As stated in Clause 2.5.1 in Chapter 12, this should not be confused with your HACCP flow diagram.

Clause 4.3.2

This is one of those clauses that has been derived from several clauses in the previous issue. In this case, most of them were moved from Section 7.

To meet this particular requirement, you must have a plan that defines access points for personnel, staff facilities and the routes they take through the site. In addition, it must show the routes for removal of waste and the movement of re-work and, of course, these shall be such that it is safe for the product. The Auditor will want to study your plan for this aspect as for the others. It is also an important part of your staff training that you emphasise the proper access points and routes, perhaps during induction.

This next part does need some explanation in that it opens with a conditional 'If it is necessary to allow access through production areas…'. This refers to the situation where people have to move through one area to get to another, it does not refer to staff who remain in their working area. Thus, for example, if staff, visitors or contractors have to move through raw materials handling to get to production or through production to get to storage, you must provide designated walkways.

The Auditor will ask if there is traffic from A to B and if so will want to see the routes designated. This might be done by barriers or indications painted on the floor or by other ways of clear signage. The Auditor will also be alert during the site tour to anyone seen not to comply with such walkways.

Staff facilities in particular should be considered. The idea is that they allow access by simple and logical routes. Thus, changing facilities toilets and staff rest rooms or canteens should be easily accessible but at the same time the protection of the product from hazards is the main issue. Therefore, do not think that a slight diversion in order to protect the product will lead to a nonconformity.

The positioning shall facilitate any necessary changes of clothing before and after using the staff facilities. So quite a lot to consider.

Example

The site plan for a bakery did not indicate flows of raw materials (ingredients, packaging), WIP (sour dough), final products, personnel, or waste materials, or recently relocated freezer store.

A nonconformity was given.

Clause 4.3.3

This requirement concerns procedures for access for all visitors. They shall be made aware of the special requirements of your site. Being a visitor, the Auditor will expect to have been informed of such requirements and any hazards, so make sure that you do this before they are allowed to enter the site. Also, the Auditor will ask how certain other visitors such as contractors, drivers are informed of your rules. Drivers in particular are a good source of material for Auditors.

Example

I was once at an old bakery watching the unloading of raw material from a vehicle through a roller door. The door was open throughout unloading and left open for some time afterwards. After watching the operation, I had walked away when I noticed out of the corner of my eye the driver wandering into the factory. He was still dressed in his outdoor clothes and, of course, no head covering. In addition (can you guess?) – he had a cigarette in his mouth.
   Several nonconformities were raised around this incident!

The clause also concerns contractors who are carrying out maintenance. They must be supervised by a nominated person. In this case, supervision would have the same meaning as it would for any member of staff. Thus, while the Auditor would not expect someone to be constantly looking over the contractor's shoulder, there needs to be some evidence that contractors are supervised. Bear in mind that contractors may not be on site when the Auditor is there, so you must have a system in place for this that the Auditor can see.

Clause 4.3.4

This clause only concerns low-risk areas and the arrangement of process flow coupled with procedures and how they must be designed to protect the product. The Auditor will be looking at all aspects of the process to see where the product is exposed to potential contamination at all low-risk stages of production.

Example

At a specialist bread maker, the Auditor found that not all ingredient storage was adequately segregated from process areas; notably palletised bagged flour and premix.
   Because they also had no effective procedures to prevent cross contamination, here a nonconformity was given.

Part Two

Should the Auditor consider that the positioning or design of the plant puts the product at risk, nonconformities may result.

Example

Many years ago, a retailer saw a practice in a pasta factory that it did not like. The company was making egg pasta and they used liquid egg from one tonne tanks, which they wheeled into the production area and dispensed into the mix by hand. The retailer asked them to install an automatic dispensing system whereby it would be pumped from the tanks in the chill room directly to the mixers.

The factory duly did this but the pipework was poorly designed and unbeknown to the factory had dead space within it such that it could not be cleaned through properly. The result was a serious *Salmonella* outbreak.

When I was last in the factory (which was some years ago), they had reverted to dispensing the liquid egg manually.

Obviously, the point is you must think about all process flows.

Clause 4.3.5

This requirement concerns high-care areas and the protection of products from microbiological contamination by means of segregation from other zones, preferably with physical barriers. All the aspects considered here in terms of product flow, staff and waste movement are important, of course, as are airflow, air quality and utilities provision (e.g. water). Note that the clause deliberately uses the word 'should' in terms of physical segregation. It does allow for the fact that physical segregation might not be possible. This might be the case in very small sites such a small sandwich maker. In this situation, you must carry out an assessment of contamination risks and come up with an effective alternative. Time separation of processes could be the answer but this must be done in a properly controlled manner which would include strict cleaning schedules and all the usual high-care requirements such as staff entry facilities (see Clause 4.8.4).

The Auditor will be looking at your general practices during the tour and this clause might also raise questions about procedures and staff training.

Clause 4.3.6

This clause concerns your high-risk areas and has similar wording to the previous clause to begin with. However, note that in this case there 'shall' be physical segregation from other areas of the site.

This requirement for physical segregation caused some companies problems when it first appeared in Issue 4. Previously, it had been possible to use procedures such as the timing of different operations to segregate low-risk from high-risk operations, especially in small companies with limited space. Others were able to operate with less than complete physical segregation provided that staff and ingredients did not cross over between high and low risk. Now, it is essential that physical segregation is complete. As well as preventing overtly physical contamination and segregating staff and equipment, you may need to take special measures, for example to ensure that possible passage of air from low risk to high risk is prevented (see also Clause 4.4.13).

The Auditor will be looking carefully at all aspects of this. Again, the site plan will be used by the Auditor to look at the process flow; however, the site tour is the most critical here, for it is only by walking through the process that the Auditor will understand (in three dimensions) the barriers that you have in place.

Examples

A manufacturer of ready meals, pies and baked desserts did not have complete segregation because of their process flow and other issues. The Auditor found that the flow of some equipment, for example 'dolly' bins went from the low-risk area through high-risk area.

Clearly, the site had problems here and a nonconformity was given.

At a manufacturer of cooked pies and pastries, the structure of the site was such that the unwrapped finished products were being taken outside on trolleys to be wheeled across a yard to the separate packing department.

The Auditor considered that this was way too risky and a nonconformity was given.

Here is another example although this was not identified during a Global Standard audit but during a special bespoke audit.

A manufacturer of cooked chicken portions had a travelling oven. Raw pieces were placed on one end and the cooked pieces taken off the other. There was no wall or other barrier between the intake and off-take ends. A member of staff who placed the raw chicken on at one end was also walking down to the far end to take samples off the line and take core temperature readings. He was not changing or hand washing between operations.

This hair-raising example would certainly have led to a major nonconformity under the Global Standard, possibly even a critical.

It is accepted that in every high-risk operation, there has to be some transfer system from low to high risk. This clause also requires that such transfer points be located and all related practices be such as to minimise the risk of product contamination. Usually, this requires structural design to facilitate this and again the Auditor will pay particular attention to transfer points during the site tour.

Part Two

> ### Example
>
> A fish manufacturer had a smoking operation producing ready-to-eat smoked fish. The Auditor noted that there was no barrier between the low-risk and high-risk ends of the smoking kiln. Products were passed to the high-risk chiller by low-risk staff.
>
>   Clearly, there were several issues going on here and the Auditor could have considered this under more than one clause. However, as the main issue was that transfer from low to high risk was not protected, a nonconformity under this clause was given.

The Auditor will be looking to see that both your factory design and your procedures take special account of this issue. As well as the evidence from the site tour, the Auditor will also want to take account of work instructions. Any related work instructions should clearly emphasise the need for vigilance at transfer points.

There is a final point to the clause, which is that you shall also employ certain practices to minimise contamination. Disinfection is given as an example here. Disinfection is sometimes used on raw materials, for example the plastic covering on blocks of raw material that are passed into a high-risk area from low-risk may be sprayed or wiped with a biocide before it is taken off. The Auditor will make a judgement on this depending on your particular situation and how well you control transfer points.

Clause 4.3.7

This requirement is about working space and how the lack of it might compromise product safety. There are many potential instances in which this might occur if your site has been poorly designed or laid out, especially where sites grow piecemeal and additional equipment is installed, perhaps as business grows.

Try to ensure that there is sufficient space around equipment. Similarly, you must have enough storage space so that items are not stored against walls or generally obstructive. Items that are stored for long periods should not be block stowed for example.

The Auditor will be looking at this aspect during the site tour. Where an Auditor finds issues with space, they may not be easily resolved. Here, the key is good planning of layouts and giving consideration to contamination hazards when doing so. In effect, this is reflected in the SOI.

Clause 4.3.8

Sometimes, it is necessary to have temporary structures in place, for instance when you are re-building the fabric or installing new plant. You may have temporary partition walls or ceilings. The Auditor will pay due regard to this but you will not be penalised provided you ensure that it does not create contamination hazards. In particular, this clause requires that it does not lead to either pest harbourage or product contamination. The Auditor will closely inspect any such structures during the site tour and will want to know how long such work is going on. You must ensure that any unfinished surfaces are well screened from any exposed product. It is a good idea to include such areas in your own fabric audits.

Clause 4.4: Building Fabric – Raw Material Handling, Preparation, Processing, Packing and Storage Areas

The subtitle of this SOI makes clear that it is about the internal fabrication of the site and that all should be suitable. The detail is in the subsequent clauses. Here, the Auditor will be looking at the condition of the site during the tour.

> ### Key points in Clause 4.4
> The importance of surfaces: walls, floors, ceilings and windows
> What to do about doors
> Sufficient, safe lighting
> Ventilation
> Do you need positive air pressure?

Clause 4.4.1

This clause concerns the design and structure of the walls. You will need to demonstrate to the Auditor that they are surfaced in such a way that they facilitate cleaning, and are reasonably smooth-surfaced so that they prevent build-up of dirt and the potential for mould growth. They also need to be maintained and kept in a good state of repair. Cladding is often a good measure. Bare brickwork is not ideal. If you are painting wall surfaces, the Auditor may ask if you have used mould-resistant paint, although such paints are not a complete answer to mould growth and are only effective if the surface is also kept free of dust and debris.

The floor/wall junction is a very important feature. It may be poorly finished off or is often the most susceptible part of a wall to damage by trucks and pallets.

Example

A manufacturer of produce was noted to have significant gaps between the base of the wall and the floor. This was very difficult to clean and a clear hygiene risk.
A nonconformity was given.

Clause 4.4.2

Floors are always a challenge in a food production area. They need a sound base structure and a good surface. The surfacing of floors has caused many issues with factories in the past where surfaces could not withstand the treatment meted out by the ingredients and/or the process. Choose your surface carefully. As you know, they are costly to replace.

The clause requires that your floors are designed to meet the demands of the process and that they are in good repair. The Auditor will be looking at your floors and will take a view on their condition relative to the nature of the products and to the perceived risk to product.

Examples

An automated bakery plant also had a department producing sandwiches. This was a properly segregated high-care operation. However, it was noted that a chiller on the low-risk side of the sandwich operation had a floor in poor condition.

At a manufacturer of low-risk chicken products, it was noted that floors and coving were generally good but some damage was noted at the intake area and in the main production hall.

Nonconformities were given in both instances.

Clause 4.4.3

Drains are a very important aspect of factory design and the Auditor will have noted them if they appear on your site plan. The Auditor will also take note of the drains during the factory tour and will be looking to see where they are sited, their routes and how this might impact on food safety.

Examples

At a fresh sausage manufacturer, the Auditor noted more than one issue relating to drainage. Waste-pipes to wash-hand basins discharged on to the floor in the production area. Similarly, the kitchen sink waste from the staff rest room discharged through the wall on to the floor in the production area.

This rather poor state of affairs resulted in a nonconformity.

At a cannery, drainage was well laid out and maintained. Drains from the on-site laboratory did not flow back into production areas. However, the mesh/screen of the drains was so big that they might have led to the chance of rodent entry.

A nonconformity was given.

The route of drainage from any laboratories on site should be planned so that it avoids production areas and in particular any possibility of contaminating the food from this route. Again, the Auditor may want to see plans of drainage to confirm this and may also question any laboratory staff on this issue.

Drains should also be well maintained. If the Auditor finds that drains are smelly, excessively dirty or actually blocked, then it will be clear that they are insufficiently maintained. In some cases, the routine microbiological swabbing of drains is a good way of monitoring this.

The Auditor will also be looking at the siting of machinery. Wastewater all over the floor is not impressive and may lead to issues being raised, especially where drainage could have been better designed or where the floor has insufficient 'fall' towards drainage. Thus, this requirement also concerns the provision of adequate falls on your floors, so that water or effluent does not pool and stagnate. The Auditor will be looking for any instances of inadequate fall during the site tour.

Examples

At a beans cannery, it was noted that there was ponding of water due to defective flooring adjacent to the retorting area.

At a jam manufacturer, water was 'squeegied' away but there were inadequate falls to drains.

A nonconformity was given to each.

(Aside from any questions of food safety, there is a personal safety issue with wet floors. See also Chapter 10: Keeping the Auditor Safe.)

Clause 4.4.4

This is a new clause for Issue 6 concerning specifically high-care or high-risk areas. As discussed in Clause 4.4.3, drainage is always important, but it is acutely important in high-care and high-risk environments. If this is relevant to you, firstly, you must have a plan of your drain system. This plan must detail the direction of flow and systems to prevent back up of wastewater. Naturally, you must be able to present this to the Auditor.

Secondly, as with all drainage, your system must not present a risk to products, thus dirty water should not be allowed to flow back into high-care or high-risk areas.

Clause 4.4.5

One of the first things, I say to a trainee Auditor is: 'Always look up'. Ceilings and overheads are a very important aspect of any Audit and a subject that the Auditor will be very conscious of. Where ceilings and overheads are poorly maintained or poorly cleaned, they are an immediate danger to exposed product. For instance, in the case of flaking paint on a ceiling, there is an obvious danger of material falling into product. Flaking surfaces are also unlikely to be clean.

Examples

There were spider webs on the ceiling over the soaking tanks and around their platform.

At an alcoholic beverages factory, it was noted that flaking paint was present on the ceiling in the blending weigh-up area, directly above open minor ingredient packages.

In a tomato cannery, the Auditor found flaking paint and mould growth noted on the barrel vault ceiling in the production area.

At a bakery, the Auditor found some gaps between roof plates and a drainage valley in the old bakery building. Also there was a gap between the wall and roof end.

A nonconformity was given in each case.

Mould growth is another danger on ceilings, where extraction and hygiene are insufficient.

Example

At a manufacturer of pastry-based confectionery, the Auditor found that there was excessive mould development on the ceiling above the syrup production tanks. This appeared to be caused by inadequate steam extraction.

At a bakery, there was considerable fresh black mould growth on the ceiling above the 'raw' ingredient chiller.

A nonconformity was given in each case.

Where a ceiling is a false one with panels, maintenance can sometimes be an issue.

Example

At a cake and doughnuts factory, it was noted that some false ceiling panels in packaging area were missing. Clearly, there was the potential for poor hygiene and contamination.

A nonconformity was given.

The design of overheads is also important and the Auditor will be looking to see if the design lends itself to maintaining good hygiene. Girders and cross beams that trap dust, electrical conduits that are exposed with wiring, also a potential dust trap, are examples of poor design.

Example

A manufacturer of specialist sugar confectionery had roof supports, which had holes in them as part of the design. The Auditor considered that these offered harbourage for pests.
    A nonconformity was given.

Clause 4.4.6

We have touched on the subject of suspended or false ceilings in Clause 4.4.5. Often, these are a very good design solution to a building where perhaps the roof does not lend itself to good hygiene. Usually, they are based on individual panels set in a grid. This can lead to problems when panels become damaged or go missing (see Clause 4.4.5), but they do not usually create problems in themselves.
    This requirement is to ensure that unless the void is sealed, it shall be accessible for routine cleaning and pest control. It is all too easy to forget about such areas and the Auditor will look to see if there is access and that you are indeed using it to maintain hygiene and pest control.

Part Two

Examples

At a sandwich manufacturer, the area above the ceiling was not pest monitored.

A manufacturer of cheese and other dairy products had a suspended ceiling above the cutting room but there was no evidence of an inspection of the area.

Nonconformities were given in both cases.

Clause 4.4.7

This clause is something of an 'old chestnut' for Auditors and the requirement has been around for all of my many years in the industry. The requirement is for opening windows to be screened against the ingress of pests. This refers to fly screens, that is screens of mesh, fine enough to prevent small insects from flying in. These are readily supplied by pest control companies, although a good on-site engineering department can produce perfectly acceptable ones.

The Auditor will keep a watchful eye on windows during the site tour. Any windows in production or related areas that are openable and are not screened will result in nonconformities. This does not often include peripheral areas such as office windows, but could well include staff rooms and toilets.

Examples

At a bakery, the Auditor noted that open, unscreened windows in toilets could be a potential issue with flying insect access to production.

In a pasta factory, there was an underground room for a silo. Inside was an openable window at the level of the external floor, which did not have adequate protection against infestation.

Nonconformities were given in both cases.

Quite often, windows at an elevated level are overlooked by manufacturers; they should not be missed by the Auditor.

Example

A manufacturer of canned tomatoes had opening windows at a high level. The Auditor noted that those behind the boules were not provided with screens.
    A nonconformity was given.

Clause 4.4.8

In previous issues, this requirement has caused concern with some factories because it required that *all* glass windows be protected against breakage. This has since been clarified and the Auditor will be concerned with any glass windows where there is the possibility of any breakage being a hazard to product integrity. A window in an office at some distance from product is unlikely to figure in this but it will be a matter of judgement and risk assessment. Happily, for many operations in modern factory units, it is not an issue because there are no windows present at all. In older-style units, there can be plenty of glass windows in production areas. Some companies have addressed this by replacing them with polycarbonate material, others have covered their glass with protective film. Glass, which is reinforced with internal wire ('Georgian wired'), is not considered safe in this regard and will need extra protection.

Examples

At a manufacturer of food ingredients, the Auditor noted that glass windows alongside holding tanks in the Gum Department were not protected.

The upper part of glass windows on the second floor in milling area was not protected.

A nonconformity was given in each case.

Remember that you should consider all areas where items might be taken into production.

Example

A produce packer had an on-site laundry where the glass windows were not protected. The possibility existed that overalls could become contaminated in the event of a breakage.
    A nonconformity was given.

Clause 4.4.9

This clause concerns doors. Some doors are particularly susceptible to damage, especially where trolleys and bins are wheeled through. Consideration should be given to reinforcing such doors. Similarly, forklift trucks and pallet trucks can cause damage, sometimes bending roller doors out of shape. Such incidents often lead to proofing problems because the doors no longer align with the floor.

External doors have always been a potential problem in food factories. As much as being a matter of good design and maintenance, they are a matter of good staff discipline. The first part of this requirement concerns external doors that are opened. There you must still have suitable pest prevention measures in place. The Auditor is unlikely to miss open doors where there appears to be no regard for pest prevention, such as fire-exit doors left open for ventilation.

Example

A produce packer had an external door in a corridor, which led to the personnel locker rooms. This was generally left open.
    A nonconformity was given.

Part Two

A roller door which is open during a delivery is another matter and provided good discipline and controls are shown the Auditor will take a pragmatic view: a factory cannot be hermetically sealed. Rapid opening and closing doors can be an excellent feature, provided they are working.

Example

At a produce supplier, the Auditor found that the pack-house external rapid action door was continually held open.
    A nonconformity was given.

A common solution to needing intake or despatch doors open for a period is to have plastic strip curtains. They are fine in themselves but there can be a tendency to rely on them rather than closing the door. This can be a problem if the strips are inadequate.

Example

A manufacturer of canned, peeled and crushed tomatoes had a despatch door protected with plastic strip curtains. However, the curtains did not close completely to floor level, and the Auditor considered that this could permit access for pests to the finished product storage area.
    A nonconformity was given.

Staff discipline can also be a problem here and it is not uncommon to see plastic strip curtains held back in some way to allow easy passage of forklift trucks or whatever. The Auditor will certainly raise this as an issue.

Example

At a tomato cannery, the Auditor noted that the plastic curtain protecting the entrance to production was not closed, and there were two other instances of doors not closed.
    A nonconformity was raised.

The second part of this clause concerns proofing. Pedestrian doors are relatively easy to proof. Here, the Auditor will look at how the doors fit with particular attention to the bottom. An old guide for Auditors is to use a pen or pencil to measure any gap under a door. The idea being that any gap larger than such a width would allow mice to enter. However, current thinking goes beyond rodents and any gap that could allow entrance of pests such as crawling insects should be proofed.

Coming back to intake and despatch doors of the roller or up-and-over type, there are other areas of concern in that the sides of the doors are sometimes not close fitting. Further, there may be a gap at the top, especially with roller doors. Finally, the most difficult are concertina-type doors, which are a challenge to make fully proofed. Thankfully, these are not often seen in modern factories, but there are plenty still out there.

Examples

Pest proofing was inadequate on two sliding doors near the defrosting/blanching areas.

A manufacturer of mayonnaise and sauces was found to have inadequate proofing at the base of two of the final product despatch bays.

At a sugar confectionery factory, the Auditor found several inadequately proofed external door bases; notably, in marshmallow packing, jelly production and packing and the liquorice process area.

Nonconformities were given in each case.

Some companies seem to go out of their way to make this difficult for themselves.

Examples

A milk dairy had a semi-permanently parked trailer, used to store polythene bottles. It was parked through an open doorway immediately adjacent to bottling hall. There was no adequate proofing surrounding the trailer.

At a cake bakery, the Auditor found that flour deliveries to a particular flour silo intake inappropriately required the external door to be held open continually.

Nonconformities were given in each case.

The final part of this clause mentions dock levellers. Dock levellers cannot realistically be proofed in the way that doors can be finished off, for example with strips at the bottom and so on, but the key here is to ensure that they are close fitting and that they are not left in a position which could allow pest ingress when not in use.

Part Two

Clause 4.4.10

This requirement is for adequate lighting in all work areas. It is doubtful that an Auditor will be carrying a light meter and the adequacy of lighting may be a subjective matter. However, a rule of thumb is whether one can see adequately to read, to operate machinery safely and to clean, especially in dark corners.

Example

I was once in a fruit cannery where the syrup room was lit by just one fluorescent strip. It was very dark indeed. I gave them a nonconformity for having inadequate lighting but still managed to find a pack of cigarettes, a bottle of soft drink and several old plastic drink cups concealed in the gloom.

Needless to say, more than one nonconformity was given in this area, one of which was for the poor lighting.

This will be a matter of judgement for the auditor but clearly certain areas should be well lit. Aside from the general requirements of a safe working environment and correct operation of processes, in particular it would be expected that locations where products are inspected should have a good intensity of light.

Clause 4.4.11

This requirement concerns all light fittings and where they constitute a potential risk to product, to be protected. You may consider a number of ways of doing this including diffusers, sleeves, coatings or, if necessary, fine metal mesh. The Auditor will be examining all your light fittings during the tour, including EFKs. Any unprotected lights or damaged diffusers will result in nonconformities.

Examples

At a tomato cannery, the Auditor found that unprotected tungsten light bulbs were used at the sight ports on the 'boules'.

A manufacturer of cooked meats had a number of EFKs in place but not all the lamps were shatterproof; notably, in Unit 1 can storage area, and Unit 1 meat reception area.

A bakery was found to have an unprotected glass light bulb near door entry into the external final product freezer.

At a manufacturer of prepared fruits and vegetables, it was found that fluorescent tubes in the newly formed raw material store at the rear were not protected.

At a manufacturer of cheeses, there was a missing diffuser to an emergency light in the hand-wash ante-chamber in the dairy.

All these resulted in nonconformities.

Despite the different ways of protecting lights, the clause also gives options where full protection is not possible. Control the situation with a glass management system. The Auditor will be looking at glass control in some detail anyway (see Clause 4.9.3) and lighting will certainly be considered when this is being evaluated. However, as far as lighting is concerned, this should really be a last resort as most light fittings can be protected in some way.

Clause 4.4.12

This requirement is for adequate ventilation. The Auditor will be looking for any signs of poor ventilation such as condensation or dust.

Examples

A manufacturer of canned tomatoes had an extraction system for ventilation in the production area. However, the Auditor could see that it was not sufficient to avoid or prevent condensation.

At a manufacturer of cooked pies and savouries, it was noted that there was a build-up of condensation in the low-risk butchery from steam off the gantry and the hygiene room.

A canner of apples was found to have insufficient extraction of steam on product fillers, causing condensation on ceilings.

All resulted in nonconformities.

Clause 4.4.13

Although it has its origins in a clause in Issue 5, this is effectively a new clause for Issue 6 and a significant one because for the first time high-risk areas must have sufficient changes of filtered air to meet the Standard. Previously, there were requirements

Part Two

concerning filtered air but the Standard did not specify that it must be present for high risk as it does now. Companies must be sure to go through Appendix 2 of the Standard and be sure of any areas that are categorised as high risk. If in doubt, go through the decision tree, there as a guide (see also Chapter 6).

In high-risk operations, the Auditor will want to see documentary evidence of filter specification and the frequency of air change. Naturally, you must also maintain your filtration system and the Auditor will look at records of filter changes and maintenance generally. There is a reference to risk assessment that should be used to determine appropriate filter sizes and the frequency of changes.

Question: What is 'sufficient' in terms of air changes?

Answer: There is no definitive answer, some have given a figure of six changes per hour but obviously it will depend on the situation. At the time of writing, the BRC may publish a guideline on the subject, so please look for that in future.

Clause 4.5: Utilities – Water, Ice, Air and Other Gases

The title now helpfully defines what is meant by 'utilities' here and the objective is that they shall be monitored to ensure product is protected from contamination.

> **Key points in Clause 4.5**
> Potable water
> Water testing: sample points frequency
> Having a plan of the water system
> Using non-potable water
> Gases

Clause 4.5.1

The requirement is to establish the principle that any water that you use in food manufacture, including preparation of primary products, or in plant cleaning, shall be potable or pose no risk to product. Furthermore, the water shall be either from your local mains supply or be suitably treated.

Potable means 'fit to drink'. In reality, it is difficult to imagine non-potable water being acceptable in most circumstances and the Auditor is unlikely to be satisfied unless it is potable, either from source, or made so by any treatment that you give it (and yet see Clause 4.5.3). The Auditor will question you on the source of your water and if from the local mains will examine the quality monitoring of such water. If the water is from a local well or borehole, the Auditor may ask to see evidence of the treatment of such water, for example chlorination, ultraviolet treatment and so on.

You will also be asked about testing of water quality, which must be done at least annually. Note that you should risk assess this to determine your actual frequency and you should think about sampling points. Failure to provide some evidence of monitoring of water quality will result in a nonconformity.

Examples

At a manufacturer of fruits and vegetables, the Auditor considered that the water analysis sampling regime was inadequate (currently, only one sample point per year).

At a poultry manufacturer, it was found that water being sprayed on to the birds in the cooling/chill room was not regularly monitored.

Water not sampled and analysed from all sources within the process area, notably, ingredient water, cleaning water and hand-washing water outlets.

A bakery making pita breads had a water-monitoring regime for some areas, but the Auditor found that the regime did not include sampling of specific sources of water used for spraying on to rolls before seeding with sesame seeds, or for cleaning purposes.

At a bakery, it was found that water samples were not taken or analysed from production area outlets for ingredient water, or hand-wash sinks, or utensil wash sinks (only from canteen).

A nonconformity was given in each case.

Remember that your monitoring may also need to include frozen water.

Example

A manufacturer of fresh and frozen poultry products was using ice as an ingredient. The Auditor found no analysis or other monitoring of the quality of the ice was undertaken.
A nonconformity was given.

Clause 4.5.2

This clause is new for Issue 6. It is a requirement for a plan of the water system as a whole. The clause itself is self-explanatory and, of course, the Auditor will ask to see it.

Note: The plan shall be used as a basis for the sampling regime discussed in Clause 4.5.1.

Clause 4.5.3

This clause is also new for Issue 6. It concerns those specific situations where non-potable water is permitted for food contact such as the washing of fish in seawater or primary washing of raw vegetables at intake before peeling and canning. This might include how the water is refreshed or re-cycled. Again, the clause is self-explanatory.

Clause 4.5.4

This clause also concerns air, gases and steam where they come into contact with food or packaging. Quite simply, you must monitor the quality of these to ensure there is no risk to food safety. The quality of steam for example may be affected by chemicals added as part of the water treatment programme for the boiler system. Such chemicals should be safe for food use. Where compressed air is used which comes into contact with product, this must be filtered. The Auditor may ask how this is achieved.

Examples

A manufacturer of muesli bars was found to have no monitoring of air quality for forced air blown directly on to cereal bars and pastries during the cooling processes.

A canner of pulses was found to have no documented monitoring or maintenance of equipment for the supply of compressed air.

Both resulted in nonconformities.

Regarding gases used for modified atmosphere packaging, the Auditor will want to know that they have been purchased against an agreed specification, which shall assure that safe products are used and that the gases are traceable.

Clause 4.6: Equipment

This SOI is an objective that that the food equipment be suitable for purpose and used correctly to minimise risk of product contamination.

Key points in Clause 4.6
Ensuring equipment is suitable
Legal requirements

The Auditor may ask how the equipment was designed but will also inspect equipment during the factory tour with this aspect in mind. There are many examples of badly designed equipment or of equipment being misused; here are a few varied examples.

Examples

At a bread bakery, it was noted that the rack covers were in fact plastic bags, which were multiple-use. It was clear to the Auditor that they dragged on the floor becoming damaged. Several tattered ones were in evidence, which were clearly contamination risks.

At a dry ingredients manufacturer, the Auditor found that hairnets were being used to cover ingredients in one room. This was clearly bad practice and not only could the hairnets be a contamination risk in themselves, the incorrect use of such an item could lead to them being used for both head covering and product covering – not an appetising prospect.

Ingredient tubs being used in the 'pre-mix area' had adhesive tape attached to the lids and in the case of the ponceau 4 R, some attached adhesive tape on the inside of the container.

There was exposed lagging on pipelines at one point in the Syrup Room and also over the Alpha Labeller.

There was some breakdown of the outer skin of the Hot Caustic Tank in the syrup room. (This had been recorded in a recent internal audit also).

A manufacturer of natural yoghurt was found to have their skimmed milk powder hydration tank lids off and stored on the floor. There is a doubling up of the potential hazard here by the misuse of the lids. They were not on the tanks, exposing the product, and they were left on the floor where they might well pick up contamination.

All these resulted in nonconformities.

Part Two

Clause 4.6.1

Equipment must be made of appropriate materials (see also Clause 4.6.2) and the design must facilitate cleaning. All these aspects should be agreed before purchase.

Equipment should be accessible for cleaning and servicing. The Auditor will be looking carefully at access to equipment during the site tour and will ask questions as to how items of equipment are cleaned. The Auditor may well ask for covers to be removed to inspect inside equipment.

Remember that this also includes small items of equipment.

Examples

A number of blue-coloured food contact surface brushes were noted as being worn and with loose bristles.

A number of hand-held contact surface scrapers in the high-care pre-pack area were noted as being damaged.

A nonconformity was given for each issue.

Clause 4.6.2

This requirement concerns the suitability of food contact materials. This will not only include plant equipment but also consider food containers, especially plastic ones. If your machinery is constructed from stainless steel, you may need to demonstrate that it is of a suitable grade. The Auditor might ask for evidence that it meets legal requirements in this regard.

Examples

No evidence available of the food-grade status of the blue plastic prep room dip baskets, green chopping boards potato dip conveyor belts, grey/yellow vegetable dip containers.

At an ingredients supplier, it could not be confirmed that the black, vegetable oil feed pipe in the rusk plant was made of materials suitable for direct contact with food materials.

A nonconformity was raised for each.

Clause 4.7: Maintenance

This is an important statement, which sets the objective for an effective maintenance programme. The related clauses detail how that is to be achieved. The objectives are to prevent product contamination and to prevent breakdowns.

Key points in Clause 4.7
Planned maintenance
Inspecting equipment at pre-determined intervals
Controlling temporary repairs
Ensuring maintenance does not create a hazard itself
Lubricants
Swarf mats

The Auditor will keep an eye on all your equipment during the site tour. Issues that can cause problems include paintwork (flaking paintwork is a common problem) loose fittings such as nuts and bolts, damage to metal or brittle plastic structures.

It is a cliché but true, that a fresh pair of eyes is always likely to see things that go unnoticed by staff who are on site every day. It is also probably true to say that this clause results in more nonconformities than most. Here are a few examples of nonconformities raised, some major, some minor.

Examples

There was a rusting, wall-mounted heater in the production area.

Frayed webbings were noted on the conveyors at depositors.

The band saw was damaged and there was no evidence of corrective action.

A potato processor passed the potatoes over roller conveyors. The foam covers to the rollers on the washer were perished and defective.

Inappropriate, fragmenting, plastic ingredient scoops.

Friable, holed lining material at the bon bon moulding/cooling exit.

Badly fragmenting/damaged plastic ingredient measuring jug in the pectin preparation area.

Flaking paint on motor above mixer.

Fraying curtain at entrance to jelly de-moulder.

Clause 4.7.1

You must have a system of *planned* maintenance. This must be documented. Here, the Auditor will be looking for a systematic approach to maintenance where jobs are scheduled in advance and recorded when they are done. The plan might be anything from a simple wall planner and job sheets, to a computer database; the point being that this is not about reactive maintenance to breakdowns but preventative action.

Considering that this requirement has not effectively changed since Issue 1 of the Standard (1998), it is surprising that so many companies are not prepared properly for this.

Examples

There is no planned maintenance for the freezer.

A manufacturer of fruit pies and muffins was found by the Auditor to have no Planned Preventative Maintenance schedule for either the externally or internally performed maintenance of critical process equipment.

A nonconformity was given for each issue.

Nonconformities may arise where a system is in place but not all the equipment has been considered.

Examples

At a wine bottler, there was a planned maintenance system but the Auditor found that it was not comprehensive; notably not including all process equipment.

At a bread bakery, there was no plan for the maintenance or inspection of the filters in the air pumps transporting flour to the point of use.

A nonconformity was given in each case.

Sometimes, items that are serviced by contractors are overlooked.

Example

At a manufacturer of burgers and other meats in buns, it was noted that planned maintenance schedules did not include equipment on external contract (notably refrigeration and gas flush).

A nonconformity was given.

Obviously having a schedule in place is one thing, but you will then also need to show evidence that you are meeting the schedule.

Example

An ice cream manufacturer had a maintenance schedule in place but the Auditor noted that the maintenance schedule and the record log for maintenance did not match.

A nonconformity was given.

You must also remember to set out the requirements for new equipment as you commission it. Again, the Auditor will ask you about any new equipment that you have installed. Commonly, this is established at the opening meeting as it gives the Auditor a picture of any changes to the plant, especially if it is not your first audit.

Therefore, if the Auditor is aware of new equipment having been installed they will check that it has been included in the maintenance programme.

> ## Example
>
> At a wine manufacturer, the Auditor found that the newly commissioned screw capping process equipment had not been included within the preventative maintenance system.
>   Thus, a nonconformity was given.

Clause 4.7.2

This requirement concerns equipment that could cause a foreign body contamination problem if damaged. Examples would be sieving or filtering equipment, optical sorting equipment and so on, which could cause contamination indirectly, or the examples previously that are more direct. In such cases, you must have equipment inspection at pre-determined intervals. Clearly, the Auditor will expect to see specific schedules and records of inspection over and above any planned maintenance schedule.

You must also consider if you have any equipment that could cause foreign body contamination itself if it breaks down. An example would be slicer blades for bread, for instance. Some systems have broken blade detectors but this type of equipment must still be inspected regularly. Here is a different example.

> ## Example
>
> At a manufacturer of catering products in sachets, it was found that there were no records to show that routine checks on the air filter to the sugar-blowing unit from silo were taking place according to schedule.
>   A nonconformity was given.

Clause 4.7.3

This requirement concerns temporary repairs and requires that you control them to ensure product safety. The Auditor will certainly notice instances of 'tape engineering' during the site tour and, where they appear hazardous to product, will give nonconformities.

Part Two

Examples

At a tomato cannery, temporary repairs were noted in the juice preparation area and one instance of defective pipe insulation at high level was present.

At a manufacturer of canned, peeled tomatoes, it was noted that the valves on the 'boules' were operated by a wooden stave attached with a wire. A more suitable, permanent mechanism was most definitely required.

At an ice cream factory, an inappropriate 'tape' temporary repair on pipework was noted directly above open product on a packing line.

At a manufacturer of potato products, an inadequate temporary repair (a clamp) was noted directly above the open product on the stir-fry multi-head weigher.

At a manufacturer of meat products, an inappropriate temporary repair (tape, cardboard, film) was noted at the end of a Unit 2 metal detector conveyor for meatballs. This was above exposed finished product awaiting packing.

Nonconformities were given in each case.

The second point in this clause is that permanent repairs should always be made as soon as possible and within defined timescales. The Auditor who sees temporary repairs will ask when a permanent solution is to be made and what controls are generally in place to ensure that what is temporary does not become permanent by default.

Examples

At a pasta factory, there was a temporary repair to stop oil dripping on to the dough mixer and no plan to effect a more permanent solution.

A manufacturer of canned tomatoes and pulses was found to have temporary repairs, with no formal procedure to effect a permanent repair.

Nonconformities were given in each case.

Essentially, the Auditor will want to see a controlled policy towards temporary repairs. It may be a good idea to record the date of such temporary repairs as a reminder of their prompt full repair. I have even seen dates written on the repairs themselves, which is quite a good idea.

Clause 4.7.4

This is an important requirement, in fact, a merging of two previous clauses. You are required to ensure that maintenance itself and the subsequent cleaning are carried out in such a way that there is no hazard to product safety. It is important that maintenance staff are accountable for the work done and the Auditor may ask to see how jobs are signed off. The Auditor may also ask about the training of engineers and maintenance staff in this regard.

The Auditor will note potential hazards caused by maintenance during the site tour.

The clause also requires that after any maintenance work, all areas and equipment are clean and free from any contamination hazards and crucially you must document this. A start-up or re-start sheet is the idea, to give accountability for the handing back of the equipment to production. It is a good discipline and the Auditor will be looking from completed records that show who has completed the maintenance and signed off the item as being clean and suitable for use by production.

Example

At a manufacturer of produce, the Auditor noted that they had a procedure for contractors to sign that all equipment and surplus material has been removed after job completion. Regrettably, this was not being followed.
A nonconformity was given.

Clause 4.7.5

Any materials such as lubricants or paints shall be suitable for food surfaces. The Auditor will look at any items discovered during the site tour to see if they are suitable and may ask to see literature or specifications for any such materials that you are currently using.

Examples

There was no evidence or data sheet confirming the food-safe status of the engineering lubricant used in the bakery.

At a produce packer, the Auditor found a branded container of mineral oil lubricant in use on the carrot packing line.

A nonconformity was given in each case.

Note: See also Clause 4.9.1.

Clause 4.7.6

Workshops are not expected to be immaculately tidy but the condition of some leaves much to be desired in terms of housekeeping and hygiene. The Auditor will take in workshops during the site tour. Certainly, they must be well maintained and not allowed to become too untidy and cluttered. Pest control must be a consideration here as with all other areas and an untidy room where the floor-wall junction is full of engineering parts will be a danger.

Specifically, the clause also requires the provision of swarf mats where the workshop opens directly into production areas. This term is not defined in the Glossary and may cause some uncertainty. A swarf mat is a special floor mat designed to remove from footwear small particles or off-cuts of metal (swarf) usually from drilling or filing. Such small pieces of metal are unwanted in a food area and this is a useful safeguard.

Example

At a large-scale bakery producing baguettes, the Auditor noted that there were no 'swarf' mats at the exit of maintenance workshop leading directly into the process area.

At a jam factory, there was no contamination prevention control from the Engineering Workshop that leads into frozen concentrate filling area, notably, no swarf mat.

A nonconformity was given in each case.

Clause 4.8: Staff Facilities

This SOI and its ten sub-clauses are all concerned with the on-site facilities provided for staff. Staff facilities are an indicator to the Auditor. They say much about you, your attitudes and your resources. They are also important for staff morale.

The objective here is that they shall be designed and operated to minimise the risk of product contamination.

Key points in Clause 4.8
Matching facilities to staff numbers
High-care and high-risk entry requirements
Hands-free taps
Requirements for toilets
Controlling smoking
Controlling staff food

Much of Clause 4.8 will be addressed by the Auditor during the site tour. Design is the key to this.

> **Example**
>
> A manufacturer of tomato paste in aseptic bags and cans had no means for staff to clean their footwear after visiting the toilet. The Auditor considered that there was a slight potential for product contamination here from this oversight.
>   Thus, a nonconformity was given.

Toilets and other facilities shall also be sufficient for the number of personnel. Again, the Auditor will check this during the tour. They will have noted the number of personnel and may also ask how many toilets you have for instance and the scale of other facilities before the tour. Naturally, you will also be expected to meet the legal requirements, which, in Europe, are that there are an adequate number of flush lavatories connected to an effective drainage system.

Clause 4.8.1

This requirement concerns designated changing facilities that are to be provided for all staff and visitors where workwear is worn. Changing must take place before entry to the relevant food areas. Do include contractors in your thinking. Regarding staff, the Auditor will examine changing facilities during the site tour. For large sites, having designated changing areas is usually no problem. For very small sites, this can be a challenge but it is one that does need to be faced. The requirement is clear, you must have designated facilities no matter how small the site.

The second part of this clause is a requirement that changing facilities shall be located to provide direct access to the production areas. This means that staff do not have to get changed, then walk outside to get to the production areas. Where this is not possible you must carry out a risk assessment and act on it accordingly, the example given in the clause is the provision of cleaning facilities for footwear.

Some older-style factories and in particular certain industries such as dairies do sometimes have a problem with this.

> **Example**
>
> A milk dairy considered certain parts of its operation to be high care; however, the Auditor found that the changing facilities did not allow staff (e.g. filler operators) access to process areas without having to go outside.
>   A nonconformity was given.

Clause 4.8.2

You must provide sufficient storage facilities for personal items. This is clearly important and the Auditor will not be impressed by changing rooms full of untidy personal items, or, even worse, personal items 'stored' in the production areas, especially near to production lines. The Auditor will be looking for such items, in particular may well open drawers and cupboards near to production lines where personal items can sometimes be found.

Examples

At a bakery, some personnel bags were inappropriately stored on top of lockers immediately outside entrance to vegetable-prepared area (the majority of personnel items where in canteen area).

At a sausage manufacturer, some personal items were noted stored with protective clothing in the lockers.

At a bakery, which also had a sandwich-making department, a mobile phone was noted in the high-risk area.

At a fresh meat factory, a staff locker was noted to contain three glass jars.

At a factory making pasta meals, a glass tumbler and crisps were noted in a locker, which was situated in the bakery area.

A nonconformity was given in each case.

Note: See also Clause 4.8.9.

Clause 4.8.3

This is a merged clause containing what was previously two separate clauses. First of all, outdoor clothing and personal items must be stored separately from workwear, *within the changing facilities*. There is no specific requirement for lockers but they are a good, tidy solution.

Example

Designated lockers are in place for workwear and personal items. However, it was noted in several lockers that personal items and workwear were being stored together.
    A nonconformity was given.

Otherwise, the Auditor will be looking for some other way of segregating the outdoor clothing. Note that meeting this clause should ensure that you meet the previous one.

The Auditor will examine changing areas during the tour, may ask to see inside some lockers and will be on the lookout for any personal items in the plant.

Examples

At a low-risk bakery, outdoor personal coats and bakery workwear were inappropriately stored together on wall pegs adjacent to the staff entry door into the production area.

At a manufacturer of ambient cakes, workwear was being stored in an outdoor clothing locker.

Both of these resulted in nonconformities.

You must also ensure that your clean and dirty overalls are segregated. The Auditor will observe how dirty overalls are handled during the site tour and if you launder on site will want to inspect that area. External laundry services often provide systems whereby the dirty overalls are posted into designated containers and this is a good system (see also Clause 7.4.4).

If you are laundering on site, the Auditor will want to see how you control these procedures.

Clause 4.8.4

This is a new clause for Issue 6 as it relates solely to high-care areas as defined in the Standard. The Auditor will expect to see designated facilities just for high care, which are separate from any facilities for low-risk staff. They should be finished to a good standard and facilitate changing. They must also allow direct entry to the high-care area once changed, with no contamination of protective clothing. The clause is very prescriptive and lists in six bullet points that you must have in place.

Note that the clause requires procedures for properly donning the clothing and that the clothing be visually distinctive. The Auditor will expect to see good changing practices such as always working from head to toe (i.e. putting on headgear first, etc.). Colour coding of overalls is a standard practice in most high-care operations and the Auditor will expect to see something of the sort to ensure that they are visually distinctive from low-risk clothing.

Concerning footwear, note that unlike the requirements for high risk (see Clause 4.8.5), shoe coverings may be permitted for visitors. You may also have a boot wash at the entrance.

Concerning hand washing, you must have provision on entry to production for this and hand disinfection. If these are in similar containers, which is often the case, make sure that they are well labelled as such as they can often be confused, especially by

Part Two

visitors. Also be aware that hand washing itself must not contaminate the protective clothing. You may need to have specific instructions on this point.

Note: Unlike the requirement for high risk (see Clause 4.8.5), the facility is not required to be at the entrance to the high-care area although you have to enter via the facility.

Clause 4.8.5

This clause concerns entry to high-risk areas and the requirements are very similar to Clause 4.8.4. The main differences being that it must be at the entrance to the high-risk area, shoe coverings are not permitted for anyone and there is no specific allowance for a boot wash for high-risk entry. While they are permitted, they may be troublesome and would need good control. Effective sanitation of the wash water would have to be demonstrated. The Auditor will pay great attention to high-risk entry during the site tour. Essentially, the footwear worn in high-risk areas must be dedicated to that area anyway.

> **Example**
>
> At a manufacturer of smoked fish products, the Auditor found that there was no designated changing facility for entry to high risk, where staff could follow appropriately specified procedures. Also, protective clothing for the high-risk area was not visually distinctive.
>   This was a significant failure to meet the requirement and a nonconformity was given.

Clause 4.8.6

This requirement is applied to all production areas, concerning hand-washing facilities. They shall be placed at access to production areas and at other locations as appropriate. There is an element of judgement for the Auditor in the phrase 'suitable and sufficient'. The degree of facilities required will depend on the size of workforce and nature of the operation. Nevertheless, the requirement is clear that at least some facilities are required at production entrances.

> **Example**
>
> At a tomato cannery, the Auditor found that there was no hand-wash station at the entrance to the production area.
>   A nonconformity was given.

The facilities provided should be as a minimum sufficient quantity of water at the right temperature. This does not necessary mean hot; in fact, water that is too hot may deter users. You should ensure that it is a comfortable temperature, neither too hot nor too cold. Liquid soap and single-use towels or air driers must be provided. Roller towels cannot be considered as single use.

Examples

At another tomato cannery, they had recently installed new hand-wash stations at the entrances to production. However, at the time of the audit, there was no water supply to them.

At yet another tomato cannery, the Auditor found that the soap and towel dispensers to the hand-wash station at the entrance to production were empty.

Nonconformities were given in each case.

A new requirement for Issue 6 is that the facilities must provide hands-free taps for all kinds of production. As always, you must also provide signs to prompt hand washing. Clearly, the Auditor will look for these features during the site tour.

Clause 4.8.7

This requirement concerns two aspects relating to staff toilets. Firstly, the location of toilet facilities shall be such that they do not open directly into food production, packing or storage areas. Ideally, at least two doors will separate any toilet from these food areas. The Auditor will check toilets during the tour to judge this.

Example

At a tomato cannery, the Auditor found that toilets were not fully segregated from the production area.
A nonconformity was given.

Problems can sometimes occur when there are two doors, which separate toilets from production but where one of them is propped open or a door-closing device is not working.

Secondly, remember that the general conditions in toilets must be hygienic. The requirement means that poor facilities such as water at the wrong temperature (too cold?), no soap or lack of hand-drying facilities will be raised as nonconformities. You must also have seen signs requesting users to wash hands.

Part Two

> **Examples**
>
> At a pizza factory, the toilets were without hand drying and washing instructions.
>
> There was no soap or hand towels in the male and female toilets.
>
> Tomato cannery again, this time the Auditor found there were no facilities for hand drying in the staff toilets.
>
> A nonconformity was given in each case.

Note: Where your only hand-washing facilities are those in toilets, then the requirements of Clause 4.8.6 also apply, that is you must have single-use towels and so on.

Clause 4.8.8

Smoking is very bad news in food production. Thankfully, these days it is nothing like the problem it once was in and around food factories. Auditors will see occasional lapses but generally smoking is rightly banned in production areas and usually it is an offence, which leads to dismissal if caught doing so. Some countries have banned smoking in the workplace, but not all. The following applies to those, which have not.

In an ideal situation, there would be no smoking at all on a food site. Some companies have successfully managed such a policy and if done well there is much to recommend it. However, this requirement recognises that to have a complete ban on smoking may not be practical and in such cases a better policy is to allow smoking in designated areas where it can be monitored. These designated areas shall be so isolated that smoke cannot reach the product. Furthermore, where you do have designated smoking areas inside the building you must provide extraction to the exterior of the building. However, unless your country's legislation still permits smoking inside such areas will be outside, commonly an open-sided shelter somewhere in the perimeter.

The Auditor will ask what your policy is and will judge how you meet the requirement. During the tour, the Auditor will check any designated smoking areas and look to see how they are managed. Plenty of ashtrays should be provided and the Auditor will certainly look at the condition of the floor. Cigarette ends on the smoking room floor are only a few steps away from being cigarette ends on the production floor. This is true even of simple shelters outside in the yard. Good ventilation may also be needed and the Auditor is likely to ask what provision there is for this.

> **Example**
>
> A pizza bakery was found to have no ashtrays and signage for the smoking point outside. A nonconformity was given.

The Auditor will also be keeping a lookout for any signs of smoking or cigarettes in the plant. Ultimately, it is staff training that is the key here and the Auditor will also look at your staff rules on smoking.

In some countries, especially in Europe, much has changed in recent years. Smoking is commonly banned in all workplace areas. At the time of writing though some countries still have no such ban and in any case even in the United Kingdom, it is permitted to smoke outside a factory building. The Auditor will want to see your policies on this and will be very closely looking for any evidence of smoking on the site. It you have a complete ban of smoking on site including the perimeter, the Auditor will certainly keep an eye open for any cigarette ends on the ground during the tour and also look for other tell-tale signs such as the smell of smoke in toilets. It has often been said that by banning smoking altogether the activity is driven into secret places, making the situation worse than a controlled policy of a designated area for smoking. However, these days that sounds trite and it is possible to have a complete ban and control it well. Many companies do this. Of course, the new laws, where they apply have made this much easier.

Example

At a peach cannery, the Auditor found cigarettes, lighter, drinks cups and personal items inappropriately present in the syrup room.
A nonconformity was given.

Clause 4.8.9

Any food brought on to the premises by your staff is a potential problem in itself and you need to minimise the chances of any mishaps such as cross contamination, infestation and so on. Thus, the requirement is that all such food is stored safely and hygienically and that none must be allowed to be taken into any food production areas.

Example

A manufacturer of pastries was evidently permitting staff to store personal food items in the freezer. 'Canteen staff' food, notably spaghetti Bolognese, was inappropriately stored with frozen raw materials.
A nonconformity was given.

The bringing of food on to the premises by staff should be a visible activity. Staff using their lockers for food storage is not an option. The Auditor will check lockers and changing rooms and be on the lookout for any food items lying around.

It may be therefore that you must provide storage facilities for such food, in which case they must be suitable. Naturally, they should be hygienic and provide safe storage and in some cases this may mean the provision of a staff refrigerator.

Example

A canner of peaches and apricots was found to have no facility available for the safe storage of staff food such as a refrigerator.
A nonconformity was given.

It is especially annoying for you if you provide facilities but the staff do not bother to use them, so a degree of monitoring of staff activities is essential.

Example

At a produce manufacturer, the Auditor found that while the manufacturer provided storage facilities for food in the staff canteen, food and drinks cups were discovered in lockers.
A minor nonconformity was given.

The clause recognises that many factories allow their staff to eat outside the building during breaks. This must be in designated areas with control of waste. Clearly, the Auditor will ask about these arrangements and might ask when your breaks are and see for themselves how well this is being controlled.

Clause 4.8.10

This is a requirement that applies only to factories where catering is supplied to the staff. This is where the company supplies meals that are cooked on site: the traditional staff canteen. If the only provision for staff is a simple rest room with food storage facilities, then you can move on to the next clause.

Commonly, these days staff canteens are provided by outside contractors and there are some large organisations that provide these services. On the other hand, some companies continue to run their own canteens. In both cases, the Auditor will be looking to see how this is controlled. In effect, a canteen serving meals is a high-risk food production area and should be treated as such. Any poor practices within it could cause problems in the factory itself. In the worst-case scenario, an outbreak of food poisoning caused by a staff canteen would be a nightmare for any food factory. Another key concern would be the potential introduction of allergens.

Therefore, the Auditor will expect to see good control of hygiene, temperature controls, segregation of cooked and raw product, allergen policies and so on.

> **Example**
>
> A manufacturer of deep-frozen pizzas had their own catering staff and facilities; however, there was no HACCP-style risk assessment and no formalised, cooked food temperature monitoring system had been established for the in-house catering facility.
>   A nonconformity was given.

In the case of the larger contractors, there are usually complete systems in place including internal auditing and in a self-run canteen this is the type of standard that you should be aiming for. However, not all contractors are so thorough.

> **Example**
>
> At a manufacturer of seafood products, the canteen was run by an outside caterer. However, the Auditor found that they could not verify that the caterer's fridge, freezer and cooking temperatures were correctly monitored and recorded.
>   A nonconformity was raised.

As with all internal auditing systems, once you have set it up, do not let it lapse.

> **Example**
>
> A large liquid milk dairy ran its own canteen. However, audits of the catering facilities had lapsed.
>   A nonconformity was given.

Clause 4.9: Physical and Chemical Product Contamination Control – Raw Material Handling, Preparation, Processing, Packing and Storage Areas

This SOI should require no special explanation. It is one of the few that was not amended for Issue 6. You are expected to prevent product contamination as far as possible.

Part Two

Key points in Clause 4.9
Control of chemicals
Taint from building materials
Control of metal
Glass and other brittle stuff
Handling glass containers
Records
Is wood ever necessary?

The Auditor will assess your procedures to that effect but the most likely source of nonconformities here will be actual risks that the Auditor sees during the site tour. Here are some examples:

Examples

Sugar and alginate were added to cream without screening through a sieve.

A 'bic' type plastic pen was in use in the dairy.

There was an unmarked bottle containing solvent on the passata line.

Inadequate (flimsy) knife in use in the syrup room.

Inadequately protected, exposed insulation material on an external wall adjacent to the bulk out-loading bay door and the big bag hopper (itself uncovered).

No product protection kick plates at the top step of the stairs crossing over open product conveyors.

Torn, wet, fragmenting cardboard used as traceability document inside crate of prawns.

Ingredient containers were not covered.

No knife blade integrity monitoring system (notably, in preparation area).

Inappropriate storage of spice mix bags immediately beneath a leaking hose in the mashed potato production area.

Inappropriate 'snappable' blade knives in use (see Clause 4.9.2.1).

There was risk of product contamination from empty product trays stored directly on the floor.

At a manufacturer of frozen pies and sausages, finished product was in trays in both the despatch area and in transit. It was noted that the top trays were always uncovered with the potential for foreign body contamination.

This list could go on but I think this illustrates the kind of things that the Auditor will pick up. They could be summarised as poor maintenance, lack of covers, lack of filtration or sieving, inappropriate equipment and poor practices. Try to see your site with a fresh pair of eyes as often as you can.

Clause 4.9.1: Chemical control

Clause 4.9.1.1

This requirement is for the control of chemicals. What is expected here is helpfully listed in the clause. Note that the seven bullet points listed are the *minimum* requirement.

Many food manufacturers buy chemicals from specialist suppliers and this is the best way to ensure that they are suitable for food use, fit for purpose and suitably labelled, as required here. Even so, the next part of this requirement is your responsibility. You must ensure that chemicals are stored hygienically to prevent product contamination.

The Auditor will ask to see copies of manufacturer's data sheets and instructions and may ask how cleaning staff are trained in their use. During the tour, the Auditor will look for any signs of unclosed, unlabelled or insecure chemicals.

Thus, the Auditor will be looking for a complete system for chemicals. The purchase of chemicals must be approved and there shall be an approved list. The Auditor will want to see the list and may ask to see safety data sheets and specifications (which confirm their suitability for food use). There must be a segregated storage area just for chemicals, preferably locked.

Examples

At a manufacturer of Chinese dumplings, the Auditor found that chemicals and mops for cleaning the factory were stored in the toilet.

A manufacturer of croissants and cakes was found to have a container of liquid detergent near to the production line and two spray oil containers on the thermoplastic filler machine.

A manufacturer of crisp breads was found to have an unlabelled clear liquid for cleaning in a spray-jet container on a packing line.

At a produce packer, cleaning chemicals were incorrectly stored above product bins.

All serious issues and a nonconformity was given in each case.

You must prevent the use of strongly scented products in the production area.

Example

Some domestic scented cleaning chemicals were in use in the bakery (notably Fairy washing-up liquid and Atlantic Fresh sanitiser).
A nonconformity was given.

Naturally, cleaning chemicals must not be left near to product.

> ## Example
>
> A jam factory was found to have inappropriate storage of cleaning chemical (Mip SP) near to the hot-water tank and another chemical on wall ledge near process area entrance door (not in the dedicated cleaning chemical storage area).
>   A nonconformity was given.

One of the requirements here is for chemicals to be identified at all times. This means the use of proper labels, especially on the containers used by staff. This is not an uncommon issue. Note that this does not refer to just cleaning chemicals.

> ## Examples
>
> At a manufacturer of canned hot dog sausages, the Auditor noted printer solvent inappropriately stored in a sunflower oil spray-jet container.
>
> Generally, good facilities were in place to prevent product contamination. However, a cleaning solution, used during line clean down, had been prepared in a food-grade container.
>
> A raw fish producer was found to have cleaning fluid on a shelf in the packing area and an unlabelled dispenser of sink cleaner in the smokehouse.
>
> During a tour of a cheese manufacturer, the Auditor noticed two unmarked bottles of sanitiser spray.
>
> At a confectionery bakery, the Auditor noted that the CIP chemical containers (for the buttermilk tank and pipe work) were not identified for contents.
>
> A nonconformity was given in each case.

Another of the minimum requirements is for their use to be by trained personnel. The Auditor may ask to see evidence of such training.

Clause 4.9.1.2

This is a new clause for Issue 6 that covers the subject of potential taint from materials such as paints, sealants, adhesives and so on. You must have procedures in place to prevent the risk of taint to product and, of course, the Auditor will need to see these documented.

Clause 4.9.2: Metal control

Clause 4.9.2.1

The first requirement here is for a documented policy concerning sharp metal implements. The Auditor will want to see a policy that is comprehensive. A good idea is to itemise and number knives, scissors and so on. The policy should enable the controlled issue of such items and the tracking of them, including their disposal out of the factory.

The Auditor will expect to see blade replacement systems for relevant equipment such that damaged or worn blades are accounted for and disposed of and that they are replaced only by authorised staff.

Examples

A few years ago, I was in a bakery that produced baguettes where staff were using old-fashioned razor blades to make the characteristic diagonal cuts in the top of the dough.

When the blade was past its useful sharpness, the operators posted it through a slot into a small cardboard box, which resided by the conveyor. There were so many issues here that I hope I need not list them all but needless to say they received a nonconformity for the lack of control of these blades.

In a manufacturer of Chinese dumplings, staff were issued with knives. The Auditor noted that there was a fortnightly blade control system but felt that this was insufficient.

Nonconformities were given.

New for Issue 6 is a requirement now to have a record of inspection of all such items and a system for investigating and recording any lost items.

The next part is straightforward: do not use snap-off blade knives. These are the type of knives, often sold for do-it-yourself purposes where a segment of blade can be snapped off as it wears.

Example

An unacceptable snap-off blade knife present in the repacking area of a bakery. Furthermore, they had no inspection or record of bakery bread slicer blade integrity.
    Two nonconformities were given.

The Auditor will want to see that where knives are necessary you are active in preventing this, so good staff training and instructions will provide suitable evidence.

Clause 4.9.2.2

This is a specific clause regarding staples and other hazardous items in packaging. Staples are not common in food packaging. They do appear in some outer cases and for example in some stringed teabags (although decreasingly). Other packaging items, which might cause product contamination include the following (not all of them metal):

- Metal tags on certain meat products.
- Sharp plastic off-cuts from the thermoform process.
- Other off-cut material.

The Auditor will be looking to see that you have taken all possible precautions against the possibility of product contamination. With regard to staples in packaging, you should avoid their use at the outset. Regarding the general use of stationery, paper clips and staples shall not be used in production areas at all.

Clause 4.9.3: Glass, brittle and hard plastic, ceramics and other similar materials

Clause 4.9.3.1

The presence of unnecessary glass or brittle materials shall be excluded from open product areas. Some glass might be necessary, for example sight glasses. In such cases, the glass must be protected.

Firstly, therefore the Auditor will want to see evidence of a risk assessment for such materials.

See also Clause 4.4.8 – windows and Clause 4.4.11 – lighting.

Clause 4.9.3.2

This is a requirement for documented procedures for the handling of glass, brittle plastic, ceramics and similar materials. It includes three minimum requirements: the first is for what is used to be referred to as a 'glass register'. In recent times, the concern has extended to all brittle material as well so this is now included. The Auditor will expect this to be comprehensive, items to think about might be as follows:

- Windows
- Light fittings
- Fly killers
- Display screens
- Sight glasses
- Machine covers
- Clocks
- Push buttons on equipment
- Electrical plugs and sockets
- Ceramic materials
- Mirrors in changing rooms

You also need to include all relevant areas, including delivery vehicles.

Examples

The Auditor raised a nonconformity at a confectionery factory because the glass/hard plastic register did not include the raw material or final product storage areas.

At a produce packer, the glass/hard plastic list was not exhaustive, for example clocks and mirrors were not included.

Nonconformities were given in each case.

The second part of the requirement is that the items on the register are checked for damage. The frequency of checks should be associated with your assessment of the potential risk of item concerned. The Auditor will need to see evidence of your checks in the form of records and some indication that the frequency has been risk assessed. Should any issues be highlighted on your checks, the Auditor will also ask about any corrective action taken.

The Auditor will also be looking at glass, hard plastic and other brittle items during the factory tour. In fact, it is likely that the Auditor will do this before looking at your register because any damaged items noted will then raise two possibilities. A nonconformity may be raised because of the damage itself, also the effectiveness of your checks may be questioned if you have not noted the damage yourself. The following is a range of examples of nonconformities raised.

Part Two

Examples

At a manufacturer of dried potato products (flakes and granules), the Auditor found no recorded audits of the glass/breakable plastics register undertaken.

At a vending product manufacturer, the glass and brittle material in the Cup Department was not listed on the register.

At a dairy, the monthly glass audit had lapsed by 2 months.

At a fresh meats manufacturer, the glass register did not include the Coldstore.

A manufacturer of mayonnaise, sauces in glass and plastic containers was found to have a plastic panel on the open PET bottle intake conveyor that was not included on the glass/plastic register. The panel was broken/holed and no breakage report had been generated.

At a manufacturer of cooked meats, the glass register audits in Unit 2 were not performed to the monthly schedule; only three audits had been performed in first 10 months of that calendar year.

There was a mirror damaged on the edge in the hand-wash facility immediately before entry to the production area.

A manufacturer of frozen pizzas had no formal requirement to check, or record the integrity of final product vehicle internal glass lights.

At a produce manufacturer, 'Bic' pens (brittle plastic) were in use in the pack house but not considered on the glass (plastic) register.

A manufacturer of food ingredients had several issues that troubled the Auditor. The EFKs were not on the glass register; there was a damaged item seen (grill plastic); a sight glass had a part missing. All this despite a recent audit.

Finally, the clause requires that the procedures include details on cleaning and replacement of such materials. There is only a small chance of the Auditor actually seeing such work going on during an audit, so it is likely that procedures and staff training will be the focus for the Auditor here.

Example

At a large plant bakery, the Auditor found that there was no formalised procedure for cleaning or replacing light fittings to prevent product contamination.
A nonconformity was given.

Clause 4.9.3.3

This requirement should not be confused with the previous one. It is specifically for written procedures for handling actual breakages, again, based on risk assessment. Here, the Auditor will expect to see a document that includes the safe containment of any broken glass or brittle material. The requirement has a list of six measures that the procedure must include.

You must quarantine and clean the affected area. The areas must then be inspected and only handed back to production when authorised to continue. You must also include any necessary changing of workwear and inspection of footwear.

Note: This clause refers to environmental brittle materials, not product containers that are now considered in Clause 4.9.3.4.

Your procedure must also specify who is to carry out the procedures and as always with such procedures there will be training implications as well.

The final part of the requirement is that you must record all breakages of the kind discussed in Clause 4.9.3.2 or indeed any that pose a risk of contamination, in an incident report. The Auditor will want to see such records and, even if there is nothing to report, see that you have made provision for such reports.

Examples

At a milk dairy, the Auditor found that while there was a logging system in place for on-line breakages, there was no log, or other mechanism in place to record the breakage of fixed glass or hard plastic.

At a manufacturer of cakes and croissants, the Auditor found that there was no process for recording incidents of glass/brittle plastic breakages.

A manufacturer of dried fruit had no reference to handling, identifying or disposal of affected food products, within the glass breakage procedure.

A manufacturer of sauces had a glass/plastic breakage procedure, which included on-line glass jars but not 'fixed glass'.

At a produce manufacturer, the Auditor considered the glass breakage procedure to be inadequate: there was no reference to dedicated cleaning equipment or management authority.

At a rice manufacturer, the Auditor considered that incident reports for breakages were lacking detail, for example more accountability was required.

Nonconformities were given in each case.

Clause 4.9.3.4: Products packed into glass or other brittle containers

This clause refers to glass jars and bottles, but will also include ceramic dishes such as those used for pates. This set of clauses is all new for Issue 6.

Clause 4.9.3.4.1

All such packaging must be segregated in storage from food items (raw materials through to finished products) and other packaging. Clearly, this will be visible during the Auditor's site tour.

Clause 4.9.3.4.2

Of all the potential foreign body complaints, it is the thought of broken glass that worries us more than most. This clause concerns the management of breakages on filling lines. For operations where breakage is a relatively frequent occurrence such as glass bottling plants, the Auditor will expect to see a standard routine where a fixed number of jars or bottles either side of the damage are removed as a matter of course as well as all the other precautions required here. This type of routine is commonly documented and posted up in the area. The seven bullet points listed in the clause provide detail of all the aspects that are requirement and most are standard procedures in such plants. Note that you must have documented instructions, which ensure all seven are put into practice. You will need dedicated equipment for cleaning up breakages and once again staff training will be very important.

Example

At a milk dairy, the glass breakage procedure did not detail requirements for handling glass breakage cleaning equipment, for affected product identification, isolation or disposal.
   A nonconformity was raised.

It may be necessary to extend this procedure to your storage facility.

Example

At a packer of oils, a warehouse operative was unaware of the procedure to be followed to clear up broken glass.
   A nonconformity was given.

Some modern high-speed filling lines have automated systems for the washing down of the lines and the clearance of glass before re-starting. In other words, the human element is removed. However, note that such systems will still have to satisfy the next clause.

Clause 4.9.3.4.3

Records for this system must include all breakages and even where no breakages have occurred. The Auditor will need to see your records and you must also be able to demonstrate that you have reviewed and trended any results.

Clause 4.9.4: Wood

Clause 4.9.4.1

This requirement concerns the elimination of wood in all areas where it could contaminate product. In some special cases, such as starch moulding of sweets the use of hardwood material is still sometimes seen. Wood is also used for maturation vessels, for example for alcoholic drinks. In such cases, you must check the condition and the Auditor will want to see evidence of such checks. However, for most purposes, alternative materials are available. The Auditor will be keeping an eye out for wooden tools or cleaning equipment, inappropriate use of wooden pallets, wooden parts of the building fabric and so on.

Examples

At a manufacturer of licorice confectionery, the Auditor found a wooden base plate over the cooker and wooden bench legs in the 'Guillotine Area'.

A manufacturer of frozen Indian meals was found to have wooden doors (chipboard) at a sink and filling point which required repair.

The condition of wood is not regularly checked to ensure it is in good condition and clean (e.g. pallets, table).

At a manufacturer of dried potato products, the Auditor found a wooden-handled, temporarily repaired broom present in the intermediate product area.

A jam manufacturer was found to be using splintering wooden boards to cover sugar bins in the process area.

At a pulses cannery, the Auditor found wooden pallets for bagged salt on top of the tanks for brine preparation.

A manufacturer of cakes and fruit pies was found by the Auditor to have inappropriate, wooden-handled knives present in the filo and fondant process areas.

A nonconformity was given in each case.

Part Two

Clause 4.10: Foreign Body Detection and Removal Equipment

This is a new series of clauses for Issue 6 to bring together the two sides of the prevention of foreign bodies and their detection. Obviously, they are different concepts and there must be a priority for prevention wherever this is possible. Some of the clauses remain from Issue 5 but several are new additions. Note that the SOI sets out the objective that the risk of product contamination be reduced or eliminated.

Key points in Clause 4.10
How this relates to the HACCP study
Location and sensitivity of equipment
Frequency of testing
How to test detectors – and how often
Requirements for filters and sieves
Must we have a metal detector?
Magnets
Optical sorting
Cleaning food containers

Clause 4.10.1: Foreign body detection and removal equipment

Clause 4.10.1.1

This new clause should be self-explanatory. When conducting your HACCP study, you should have considered sources of foreign bodies in your process and thereby any equipment necessary to detect and/or remove them. The seven bullet points in the clause give examples of those to be considered. The Auditor will need to see your documented assessment.

Clause 4.10.1.2

To satisfy this clause, your documented system must specify three elements of removal and detection equipment. These are the type of equipment, the location of it and the sensitivity. You are expected to use industry best practice in the specification of the equipment and no doubt you will use advice from your equipment suppliers. This will take into account your specific industry factors such as product density, pack size, conductivity and so on.

Note: You must be able to demonstrate validation and justification regarding the siting of equipment (or any other factor that affects sensitivity). The Auditor will have this in mind during the site tour, but you may have to have some documentary evidence available for this.

Clause 4.10.1.3

This clause is about the testing frequency of the testing of both detection and removal equipment. Thus, we will be looking at both filters or sieves and metal detectors and other types of detection equipment.

Note: The testing frequency for detection equipment is also part of Clause 4.10.3.4, so you will need to have both in your mind as you address this.

Here, the issues are two-fold. The general requirement for foreign body detection and removal equipment is that it is tested and that the testing takes on board any customer requirements. So, for example, your client might require testing every hour. The second part is about your ability to isolate and re-test product should the equipment be shown to be out of order during a routine test. This is an interesting point because in effect it is saying that if you are prepared to isolate a large amount of product and re-test it, then your frequency of testing could be relatively low. However, you can see that it is beneficial and cost effective to have frequent testing. Also, customers are unlikely to be impressed if you only test once a day on a busy line.

Clause 4.10.1.4

When you successfully remove or detect foreign material, you are required to investigate the source. The Auditor will be looking for systems in place for this. You would be expected to record such incidents and your records should also be used to look for trends in foreign material. Any trends shown up shall be used to strengthen your prevention methods.

Clause 4.10.2: Filters and sieves

There are two clauses here. Note that the first one concerns only filters and sieves that control foreign material, not those that are used in normal processing such as the removal of haze from a beverage.

Clause 4.10.2.1

This is another new requirement for Issue 6. For this, the Auditor will expect to see that any filters or sieves that you have are properly and sensibly specified. You also need to demonstrate a system for checking and recording the material detected or removed by them. Your system should ensure that the information is used to identify specific contamination risks. If you continue to pick up certain material, then you know you have a problem that you should be able to isolate and prevent.

Clause 4.10.2.2

This clause concerns all filters and sieves and requires that you carry out inspections of filters and sieves. The Auditor will be looking for evidence that you are doing this and as required you must record checks and any investigation required of any issues.

The requirement is to regularly inspect and properly maintain such items at a frequency based on risk. The Auditor will expect the frequencies to be based on the nature of the item and might ask to see your documented frequency and evidence of a risk assessment here. Of course, this will also feature during the site tour and the Auditor might ask to look at filters themselves.

Part Two

Examples

No documented sieve mesh integrity checks of the large circular sieve immediately prior to nugget packing.

A cheese manufacturer had no formalised system for monitoring or recording the sieve mesh integrity of the newly installed salt sieve mesh located on the cheese incline conveyor immediately pre-filling of the cheese moulds.

A nonconformity was given in each case.

Clause 4.10.3: Metal detectors and X-ray equipment

Clause 4.10.3.1

This is an important clause. One of the most debated areas of food safety is whether a site must have metal detection. Is it not the truth that metal detection is a safety net that we use because we think that no matter what procedures are in place, there is always a chance that a piece of metal might contaminate a product? Or, do we do it because it gives us a 'due diligence' defence when we do have a metal complaint? Either way, the Standard does not have an absolute requirement for metal detection.

According to the Standard, you must have metal detection *unless* you are able to justify by risk assessment that it is not necessary in that it does not improve product protection. Indeed much of the food industry sees metal detection as the norm and does not question the need for it. Some manufacturers are uncertain as to whether they need it however, while some are certain that they do not. Wherever metal detectors are not used, you must justify this and this must be documented. In such cases, the Auditor will ask to see your written justification. This will need to be convincing.

For example, if you are cutting product or sieving or mixing using metal mesh or blades, it is a 'given' that you will need metal detection. The conditions where you may not need detection are few. They include industries such as liquid processing and filling where the product is fine filtered and filled without being exposed. Thus, wine is an example; also soft drinks or milk, depending on the filling systems. Naturally, if you are using an alternative method such as X-ray, you can easily justify not using metal detection as well.

Examples

No justification for breadcrumbs manufactured on site not currently being metal detected.

At a manufacturer of food ingredients, the Auditor found no justification or documented risk assessment to confirm that metal detection was not required for the paste products, which are not sieved or filtered.

At a dairy, there was no documented justification as to why bottled milk and bulk products are not metal detected.

A sandwich manufacturer had concluded that there was no need for metal detection on the basis of having no metal complaints in their history. However, because they were cutting tomatoes and other items by machine, the Auditor was unhappy with this conclusion and considered that metal detection was necessary.

A nonconformity was given in all cases.

It is usual for metal or foreign body detection, where it is present, to be referenced in HACCP and to be considered as a CCP and it is recommended that you do so.

Clearly, where you have concluded that metal detection is required you must be consistent.

> ### Example
>
> A processor of poultry produced portions by machine cutting. The Auditor found that while there was metal detection on some lines, not all final products were metal detected; notably, gas-flushed fresh chicken portions.
>   A nonconformity was given.

Note: The BRC have an excellent guideline published on Foreign Body Detection.

Clause 4.10.3.2

This clause which concerns the siting of metal or X-ray and is quite prescriptive. You must site the equipment at the latest practical step and after packing if possible. The sensitivity will be greater if on packs before going into outer cases, so better to have them after packing but before putting into cases, although this is not always possible.

> ### Example
>
> At a manufacturer of poultry products, it was found that metal detection was not situated to minimise finished product contamination. It was notably sited at the cutting/portioning stage only, that is not for processed sausage, which are produced later.
>   A nonconformity was given.

Of course, there are some products where it is not possible to detect metal in final product, for example canned foods. As it happens, in recent years, number of canneries have put metal detection units over product inspection belts before filling, just as they also may have magnets at such points. These may be useful and the Auditor will expect to see the normal procedures in place for them (see Clause 4.10.3.3).

Clause 4.10.3.3

This requirement lists three possible systems for the rejection and segregation of product where metal or foreign body has been detected. They are alternative ways of detecting and rejecting and their suitability depends on the nature of the product. They are automatic, belt stop or segregation for in-line detectors. Belt stop systems for rejecting product may be acceptable but only for large packs or for other products if automatic rejection is unsuitable such as a fragile product.

   Note: At the time of writing the BRC is considering allowing other types of products beyond large packs and fragile products.

   For this requirement, the Auditor will want to see a full check being carried out during the factory tour. This will enable them to see the reject or segregation system in action. Regarding reject systems, it is vital that the right pack is rejected, so the timing and synchronisation of reject mechanisms is very important. Malfunctioning reject arms, excessively fast conveyors or too much product going through can result in the affected (and detected) pack not being rejected properly.

Example

At a processor of dried vine fruits, the Auditor found that the rejection mechanism on the small automated bag packing line metal detector was ineffective, as it did not always push the product into the reject bin.
   A nonconformity was given.

Where automatic rejection devices are used, the clause specifically requires diversion into a secure unit accessible only to authorised personnel. Often, this is a locked box. In the early days of the Standard, this point often resulted in nonconformities and some still seem to miss this requirement.

Example

At a cheese manufacturer, the Auditor found that the line 3 metal detection reject box was unlocked.
   A nonconformity was given.

Clause 4.10.3.4

This should be a straightforward requirement for procedures for routine testing (and calibration) of detectors. The clause lists four points that the procedures must include.

   The Auditor will expect to see documents showing who is responsible for the testing. Remember, if detection is a CCP, this may also lead to questions about training. The document must also include reference to the sensitivity of equipment including any necessary variation caused by the product. The Auditor will also need to see the method and frequency of checking the systems.

Example

The metal detectors at a bakery were not being tested hourly (as stated in test procedure) but currently only three times per day.
   A nonconformity was given.

Part Two

Hourly is often considered the norm and you might also be advised on testing before and after production shifts. However, remember the requirements of Clause 4.10.1.3 and that the frequency of testing must take account of customer requirements and your ability to isolate product should the detector fail. Testing and reporting should also include the findings of the working of any alarm or reject mechanism. In the case of metal detection, the Auditor will expect to see testing normally carried out with three metals: ferrous, non-ferrous and stainless steel. Exceptions to this will have to be justified. Clause 4.10.3.5 deals with the detail of the checking procedures.

Clause 4.10.3.5

This is a new clause for Issue 6 of the Standard but there should be nothing of concern here. It is a good example of the more prescriptive nature of many clauses and this one spells out exactly what is required of the checking procedures. There are four bullet points that all checking procedures must include as a minimum. Naturally, there is a requirement that you test metal detectors with test pieces. Note that this must include non-ferrous and stainless steel unless you are packing in foil. The test must also include the reset function of the detector such that it must be able to pick up repeated pieces of metal in consecutive packs.

The Auditor will also observe testing during the site tour.

Examples

At a large plant bakery, the following issues were found:

Line 2 metal detector in the bread plant did not activate the 3.0-mm non-Fe test piece.

A stainless-steel test piece was only in use on the high-risk line.

The test pack used for the metal detector test on line 2 was not made up correctly and not passed through the detector before the test piece was inserted.

At a manufacturer of cakes and muffins, the Auditor found that for the observed test of the metal detector on line 3, the reject mechanism repeatedly failed to reject any products (with test pieces) while no corrective action was taken.

A cocoa manufacturer was found to have a metal detector tested during the audit, which was found not to be operational. The failure had not been documented on the process sheet.

At an ice cream products manufacturer, the Auditor found an inadequate rejection mechanism on line 11a metal detector; notably product was not always rejected into the locked reject cage due to a misaligned compressed air jet.

At a potato granule factory, it was found that not all metal detectors were tested with stainless-steel test pieces. This was of special concern on packing lines where finally the products pass through in-line stainless-steel sieves of 2–8 mm.

Part Two

During the audit test of the metal detectors in a bakery, the shortbread detector failed to detect the ferrous or non-ferrous test pieces when the test pieces were placed inside the boxed product (to simulate actual contaminated shortbreads).

At a chilled and frozen fish products manufacturer, the Auditor found that the line operative metal detector test did not ensure that the test pack would reject in the bin successfully.

A confectionery supplier had some good practice in that they had proper test packs made up for certain products; however, test packs (test piece buried within product pack) had not been created for the 170-g- or 3-kg-bagged product.

At a processed cheese factory, it was found that an inappropriate stainless-steel test piece was in use on the metal detector; notably 7 mm in use while a 5-mm critical limit was defined in the HACCP study.

At a sauce manufacturer, the Auditor found that the metal detector test was never performed at the full conveyor speed with three consecutive test pieces to ensure effectiveness of rejection mechanism.

A jam factory was found to have no ferrous or non-ferrous testing undertaken on either the post-cooking or post-filling metal detectors. An un-calibrated test rod (of indeterminate size) was used to test the post-cooker metal detector. A non-ferrous test piece observed only to be detected intermittently on post-filling metal detector.

Nonconformities were given in each case.

The Auditor will question you on where you place test pieces, for example inside test packs.

Example

A cheese maker had no documented evidence that the Regato cheese metal detector was able to detect the Fe/NFe/SS test pieces when inserted into the centre of the cheese (to accurately simulate metal product contamination). When this was tested during the audit, the detector was unable to consistently detect the 2.5-mm Fe test piece (failed 3 out of 4 attempts). The current site test procedure involves passing test pieces through the detector inside plastic bottles.
    A nonconformity was given.

Clause 4.10.3.6

If it is found that a unit fails a test, you must have corrective action and reporting procedures. It is likely that the corrective action will require some adjustment or re-setting of the machine and this must be fully reported.

> ## Example
>
> A cake and croissant bakery was found to have no procedure for corrective actions in the case of failure of the metal detector during routine monitoring.
>   A nonconformity was given.

The clause further requires that where a detector has failed a test, the procedures require that product passed through the detector since the last successful test be quarantined and re-tested (using a reliable detector of course).

The Auditor must see evidence of these procedures, so it is advisable that these be documented.

> ## Examples
>
> An ice cream manufacturer had no procedure for metal detection as to what to do if the machine fails for example recheck and quarantine.
>
> At a dry food ingredients supplier, the Auditor found inconsistencies in the performance of metal detector testing with varying results. The lack of documented procedures had clearly led to this situation.
>
> A manufacturer of maize flour was found to have basic corrective action instructions in the HACCP for metal detector test failure. However, these were brief and staff had not been fully trained in it, nor had it been documented as a work instruction.
>
> The HACCP plan included metal detection, however, referring to the actions undertaken on a certain day products between 16:00 and 17:00 were blocked and assessed. But, the metal detector malfunctioned at 16:00 and the last satisfactory check was performed on 15:03. It was thus concluded that relevant personnel did not understand the right procedures to be followed.
>
> At an ingredients manufacturer, it was found that there was no documented metal detection procedure describing re-inspection of final products following metal detector failure.
>
> Nonconformities were given in each case.

Part Two

Clause 4.10.4: Magnets

Clause 4.10.4.1

This clause is completely new for Issue 6 of the Standard and concerns magnets and what you should be checking. Firstly, you must document the type, strength and location of all magnets that you use. One way would be to include them on one of your site plans; however, you may not be able to include all the information this way.

Secondly, you must also have procedures in place (documented, of course) for the inspection of magnets and for their cleaning, strength testing and integrity checks. Regarding strength testing, some companies use Gauss meters that actually check the magnetic field of a magnet. It is also possible to check the pull strength of a magnet, measured in kilos. The Auditor will want to see your documented procedures for all these aspects.

Clause 4.10.5: Optical sorting equipment

Clause 4.10.5.1

This is also a new clause for Issue 6 and a very simple requirement concerning the checks recommended by the manufacturer of your optical sorting equipment. Their recommended checks shall be followed and, of course, they shall be documented.

Clause 4.10.6: Container cleanliness – glass jars, cans and other rigid containers

Clause 4.10.6.1

This clause concerns the protection of empty food containers by various means: covering conveyors, container inversion and methods of cleaning food containers before filling. Typically, this is seen in canning or bottling where the containers are flushed out either by compressed air, steam or water. They may be rotated by 90° or inverted as part of this process. The requirement is that such processes are implemented based on risk assessment, that is where necessary. The Auditor will look at these aspects during the site tour.

Examples

The can conveyor from the loading point to the filler was not covered.

At a bakery, there was no physical contamination protection for the bread dough recently deposited into the bread tins as they pass beneath the bread plant panner moulders.

In a cannery, the empty cans were being inverted before filling, but not cleaned with compressed air as stated in their HACCP system.

It was noted that purging (cleaning) of inverted cans was not present on all lines.

There were no protective covers over open, un-lidded final products or open unfilled jars/pots/pouches on Line C.

At a peach cannery, the Auditor found that at the can inverter prior to filling, there was inadequate steam supply to clean/remove foreign bodies from cans in the new factory. Indeed, there was no evidence of satisfactory steam supply operation available.

All these led to nonconformities.

Clause 4.10.6.2

This clause is new for Issue 6. Where container-cleaning equipment is in place, you must check it during production. You could perhaps check the equipment by using a standard contaminant. In some situations, there is automatic rejection, for example with some glass bottling lines. In such cases you must also check the detect/reject mechanism using test bottles. Part of the requirement is to record your checks. The Auditor will want to see bottle checks being carried out and your records.

Clause 4.11: Housekeeping and Hygiene (Fundamental)

'Cleanliness is next to godliness' is a phrase whose origins are in our distant past. Whoever first wrote those immortal words was not thinking of food factories, but for those who have been involved with inspecting food manufacture for any length of time, the concept of proper hygiene is buried deep within our psyches. Thus, it should be self-evident why this SOI is a *Fundamental Requirement* and why it must be one of the cornerstones of your thoughts and your systems.

> ### Key points in Clause 4.11
> Procedures and records
> How to define what is acceptable cleaning
> Checking cleaning before release back into production
> CIP: design and control

The Auditor will take some time to look at all your cleaning procedures and schedules as well as looking closely at all aspects of hygiene during the site tour. Therefore, you will need to have both a clean site, good systems and good documentation. The Auditor should be realistic in the sense that 'working debris', that is the natural spillages that occur within the working shift, is an accepted part of production.

Remember that significant problems with hygiene will lead to a major nonconformity against a Fundamental Requirement and hence no certification. An example of this was given in Chapter 4. Isolated incidents may lead to minor nonconformities.

Examples

A frozen pizza factory had bulk flour silos. Here, the Auditor found inadequate cleaning within a bulk flour silo intake cabinet; notably hardened and mouldy flour deposits.

Surfaces below utensil sink dirty and encrusted with debris.

At a manufacturer of potato-based products, the Auditor found inadequate deep cleaning of a carrot peeler out-feed and the potato post-dipping conveyor belt.

All were given minor nonconformities.

Part Two

Clause 4.11.1

It is essential that cleaning is a documented activity, it is the only way to ensure that all necessary jobs are done properly and at an appropriate regularity. The Auditor will be looking for you to meet this requirement by having documented cleaning procedures. The clause lists seven *minimum* requirements that must be in your procedures, so ensure that you include these in your procedures.

Example

A company which had both a bakery and a produce preparation area was found to have no detailed documented cleaning procedures for the bakery area, for example mixers, provers, ovens, bread slicer and so on. Also, there were no detailed documented cleaning procedures for the vegetable preparation area, for example dicers, slicers and peelers.

A nonconformity was raised.

One key point is that you must specify who is responsible for cleaning. This may be a special cleaning crew, or the operators or a combination of the two. In the last case, it would be particularly important to set out exactly who is responsible for which task.

Having looked at your site plans and toured the site, the Auditor will want to see that you have included all necessary items and locations, including peripheral areas such as outbuildings and storage areas and all relevant items within the factory.

Examples

A manufacturer of chocolate cakes had a cleaning schedule but the Auditor found that it did not include overheads and high-level equipment.

A manufacturer of hamburgers and sausages had a cleaning schedule but the Auditor found that it did not include drains or ceilings.

Both were given nonconformities.

The frequency of cleaning and the cleaning methods must also be specified in procedures. Cleaning methods may be given to you by your chemical supply company but you may have to write your own versions depending on the nature of your site. It is important that the operators understand correct dosing and rinsing for example.

Part Two

Examples

At a manufacturer of grated cheese, the Auditor found no formal control of cleaning chemical dosing.

A high-risk plant was found to have documented cleaning procedures in place but the Auditor considered them to be insufficiently detailed, for example they did not give chemical dosage rates.

The cleaning form does not contain a rinse step after lysoform sanitising.

A nonconformity was given in each case.

The next stage is to ensure that staff carry out the tasks in accordance with your procedures. The Auditor will investigate this perhaps by asking cleaning staff how they know what to do and by seeing the effectiveness of cleaning during the tour.

Examples

At a manufacturer of Chinese meals, an excessive build-up of debris on a dough mixer led the Auditor to conclude that the cleaning procedure for this machine had not been followed completely.

At a bakery, the Auditor noted that the dedicated colour-coded brushes were not being used as per the procedure.

A nonconformity was given in each case.

Equally, the use of the correct chemical is essential. Staff should be instructed not to mix different cleaning chemicals, a practice that can be hazardous to the staff. Your procedures shall also include reference to the correct cleaning materials to be used. This might involve colour-coded equipment for example.

The cleaning process must be thorough and controlled to ensure that it does not become a hazard in itself. Cleaning must be done, so that debris is not merely moved from one spot to another where it might be a hazard. The use of certain chemicals must be controlled. The Auditor will look at your procedures and check on how cleaning is done during the tour.

Finally, there is a requirement for records of cleaning. This should be self-explanatory and must be set out in your procedures as well and this must include a statement of who is responsible for cleaning verification. Almost universally, cleaning records are kept in the form of cleaning schedules, which are signed off by the operatives. The Auditor will be looking for this type of thing but will closely look at dates to see if the records are complete.

Examples

At a manufacturer of fresh and frozen meat products, the Auditor noted that there were no records of deep cleaning of the band saw.

At a bakery, there were no records maintained of monthly cleaning tasks, for example ovens.

A specialist bread bakery had cleaning records for daily and weekly tasks but had no records maintained for monthly cleaning tasks.

Nonconformities were given in each case.

Clause 4.11.2

A completely revised clause for Issue 6, this requires you to define what is acceptable in cleaning performance and this has to be something measurable even if it is visual. Thus, you will need to define what is required of cleaning even if it is a visual measure. Other possibilities (perhaps easier to define) are microbiological levels or ATP (adenosine triphosphate) techniques are given in the clause. The Auditor will need to see this documentation.

Once you have set your levels of acceptable cleaning, you must then validate your cleaning methods such that they can achieve your targets. Validation of procedures does not mean that you have to do this every time that you clean; however, it is likely that you will want to check the efficiency of cleaning reasonably frequently. The Auditor will need to see records of your cleaning validation.

Clause 4.11.3

As pointed out in Clause 4.11: Housekeeping and Hygiene (Fundamental), hygiene is a vital part of your systems and must be adequately resourced. Cleaning must be as deep as possible and in particular it is most often necessary to dismantle equipment in order to clean it properly. This may involve engineering staff as well as cleaning staff so needs to be properly scheduled to involve the necessary personnel. Also it will generally only be done outside production times. The Auditor will be looking for documentation or some other evidence of scheduling of such times. Cleaning itself should not be a contamination hazard.

> ### Example
>
> Cleaning of utensils in the high-care pre-pack area was noted as taking place during production and the cleaning facilities are adjacent to the packing line 1.
>   A nonconformity was given.

This is a requirement that clearly states that you must have staff training. Where you must open or dismantle equipment, it may be necessary to train specifically for this. The Auditor will want to know who is responsible for cleaning and ask to see appropriate training records. Note that this could also apply to any cleaning contractors that you use and it might be a good idea to have some information on your contractor's systems of training.

Clause 4.11.4

It is not enough to set out detailed hygiene procedures and to record cleaning, you also have to verify that cleaning has been carried out properly. This clause is about verifying cleaning and in effect positively releasing equipment back into production. The Auditor will want to see evidence that someone, preferably not the cleaner, verifies the cleaning and records this verification. There are different levels of verification possible here. In some cases, such as high-care or high-risk areas verification of cleaning might involve some microbiological swab testing or residue testing of the ATP-type as mentioned in Clause 4.11.2.

Any issues that are shown by this verification should result in improvement or corrective actions. The Auditor will look closely at your records and follow any poor results and ask to see corrective action records. If they can be recorded on the same sheet, all well and good. If not, have your other relevant paperwork available.

The following examples resulted in nonconformities.

> ### Examples
>
> A produce packer had Cleaning Effectiveness reports that showed regular bad results (honest at least), for example for the floor and protective clothing. Preventive action was not effective.
>
> At a cooked meats factory, the post-production (daily) cleaning audit sheet for a particular day had a low score (1). There was no corrective action evident.

Part Two

Clause 4.11.5

This clause requires some detail on your cleaning equipment.

Cleaning chemicals and equipment shall be suitably identified, for example colour coded and stored hygienically.

Examples

At a produce manufacturer, the Auditor found that colour coding of cleaning utensils was random. Thus, it was possible that utensils used in the toilets could be used in production.

Cleaning utensils are colour coded by area instead of by food contact/non-food contact segregation.

There was no clear distinction between food contact and non-food contact cleaning utensils at a bakery.

Nonconformities were given.

Clause 4.11.6: Clean in place (CIP)

There is much more focus on CIP in Issue 6 than previously. Modern CIP systems are specialised pieces of equipment, which have good safety systems in-built.

Clause 4.11.6.1

CIP systems must be properly monitored and maintained; their function and structure must be correct. These are very important considerations. Because of the automatic nature of some CIP systems it is easy to assume that they are working well, but monitoring is essential. This is particularly true of rinsing systems that must ensure that all cleaning chemical is removed before food is passed through.

Examples

At a dairy, there were no records maintained of the in-house pH check of the buttermilk silo post-CIP rinse waters.

At a wine manufacturer, the Auditor could find no evidence of post-chemical CIP cleaning rinse water checks.

A nonconformity was given in each case.

Some CIP systems have a partially manual aspect.

Example

At a tomato cannery, the description of the manual CIP system was not detailed enough (e.g. tomato paste aspiration implant, evaporator implant, tomatoes and beans pasteurisers and soaking tanks).
  A nonconformity was given.

It is also required that the CIP system itself is suitably segregated from production because clearly they often use very toxic or corrosive chemicals. The siting of your system and reservoirs of chemicals is therefore important and the Auditor will look at this during the tour.

Clause 4.11.6.2

This clause and Clause 4.11.6.3 are new for Issue 6. This one is quite detailed and prescriptive and as such is largely self-explanatory. Essentially, there are three parts to be aware of. Firstly, you must have a schematic plan of your system so the Auditor should ask to see that.

Secondly, you must have an inspection report or some other form of verification concerning the four bullet points in the clause. The bullet points all concern the design and functioning of the system to ensure that it:

- cleans effectively and
- ensures that the cleaning chemicals do not contaminate the product.

You may wish to consult your equipment supplier or a specialist contractor on this
Thirdly, if you make any alterations to the system, it must be revalidated.
Note: A double seat valve (fourth point) is sometimes called a block and bleed valve.

Clause 4.11.6.3

As with the previous clause (Clause 4.11.6.2), this one is very prescriptive and therefore largely self-explanatory. Although it is colour coded to indicate that it will be mainly part of the site audit, the Auditor will need to see some documentation such as records of validation of the process conditions (first bullet point); similarly, records of analysis of rinse waters (third bullet point). Note that for the first bullet point there are up to four measures to define and validate. Again, it might be wise to involve your contractor in this exercise at the outset.

Clause 4.12: Waste/Waste Disposal

The SOI requires management of waste disposal with the objective to prevent accumulation of waste, attraction of pests and therefore risk of product contamination.

Remember that this will include waste methods inside the factory as well as outside.

Key points in Clause 4.12
Correct maintenance of waste facilities
Animal waste
Animal feed
Disposing of branded products

The Auditor will want to see waste being tidily disposed of and not being allowed to accumulate for too long before removal. Waste bags or bins in good condition which are themselves kept hygienic should be the norm. It is not always necessary for internal waste bins to be lidded. Waste should be collected in designated containers, which are not used for anything else.

Examples

A confectionery manufacturer was found to have some ingredient sacks inappropriately used for waste materials.

At a sausage manufacturer, the Auditor found that suitable and sufficient receptacles were not provided for waste in the production area.

At a manufacturer of pies and pastries, the unpleasant smell from a waste bin near the pie stamping line indicated that it had not been emptied for some time.

Nonconformities were given in each case.

Clause 4.12.1

This clause requires that appropriate waste be categorised and removed by licensed contractors only. Thus, animal waste for example must be segregated from non-animal waste. In the United Kingdom, the means of disposal of animal waste is very strict according to legislation (The Animal By-Product Regulations, 2005). Environmental issues as well as food safety should also be a consideration. Other forms of waste are also likely to be restricted to being collected by licensed contractors.

Records of disposal must be maintained and the Auditor will want to look at these. Although this clause is colour-coded green, indicating a certain amount of document review, the Auditor will certainly look at waste storage in your perimeter.

Clause 4.12.2

This is a new clause for Issue 6 and it concerns animal feed. This must be segregated from food waste and properly managed.

Example

Unprotected/uncovered external pig-food waste trailer noted.
   A nonconformity was given.

Clause 4.12.3

This clause concerns the external management of waste. The Auditor will take this in during the site tour and will hope to see well-managed bins, skips and containers. These shall be closed or covered to avoid attracting vermin. The area surrounding such containers should also be well managed. There are five specific requirements listed including that they be well maintained, covered as appropriate and emptied sufficiently often.

Examples

A milk dairy was found to have an uncovered, externally sited waste container (used for packaging and milk). In this case, the Auditor considered that this was not sufficiently managed.

At a food ingredient factory, the Auditor found external waste containers to be inadequate on the day of the audit. The plastics container was overflowing; food waste bins were not lidded (but emptied daily). The paper container doors were open.

Nonconformities were given in each case.

External waste containers must also be clearly identified and designed for ease of use and effective cleaning. The containers will need to be cleaned and the Auditor might ask how you manage the cleaning of them and the area where they are sited.

Clause 4.12.4

This clause concerns the disposal of branded or trademarked products or any unsafe product. Customers such as retailers will not want their products disposed of in such a way that they could find their way back into circulation. Therefore, in such instances where it is not possible to remove branding before disposal and where you are using a third party for disposal, you must ensure that the third party is in the business of secure disposal. Such a third party must not only be conversant with the requirements for security here, they must also provide records of secure disposal, which include the quantity disposed.

Part Two

> **Examples**
>
> At a bakery, there was no provision for the special treatment of trademarked waste.
>
> At a manufacturer of sliced cooked meats, the Auditor found that procedures did not specify the arrangements for disposal of trademarked materials.
>
> A nonconformity was given in each case.

If you have such situations, the Auditor will ask for details about the contractor and for any records proving secure disposal.

> **Example**
>
> A manufacturer of jams had disposed of some branded product but could not provide the Auditor with any records received from the licensed waste disposal contractor confirming the destruction of trademarked final products.
>   A nonconformity was given.

Clause 4.13: Pest Control

This is a very important section that is supported by a BRC Guideline. The SOI expresses that the whole site must have an effective system to minimise the risk of pest infestation. This SOI is also particular in that it recognises the need to rapidly respond to any issues, which can obviously happen with a sudden pest infestation. So, the two key objectives here are prevention and rapid response.

Key points in Clause 4.13
Can we do our own pest control?
Site plan and other documents
Responsibilities
Where can we put toxic baits?
Correct siting of devices
Records and actions
In-depth surveys
Data analysis

Thus, housekeeping, proofing and so forth shall be a part of the culture. Pest control must be covering all areas of the site.

Example

At a bakery with a bulk flour handling system, the Auditor found the pest control system did not take account of the new prover or the flour silos.
A nonconformity was given.

Anyone with experience of pest control will tell you that good hygiene is the foundation of it. Good storage conditions are especially important in preventing stored product pests. The Auditor will inspect all storage areas and be looking for well-ordered areas with space between product and walls. The facilities should be dry with walls, floors and roof in good condition. There should be no gaps in the structure and the discipline on door closing must be good.

Examples

At a produce packer, the Auditor found that the packaging store was congested resulting in inadequate pest control access.

A Chinese meals manufacturer was found to have packaging stored in the roof space. It was accessed by a stepladder and was not adequately proofed due to open windows.

A nonconformity was given in each case.

For certain products such as dried fruits, bulk flour and other cereals, the Auditor may ask for evidence of cleaning and fumigation regimes. Bulk flour is especially susceptible to infestation from flour moth and flour beetle and the Auditor will look for tell-tale signs of infestation in silo areas, sieve tailings bags, dust extractors and so on. Your pest control records will be examined for evidence of fumigation in such cases.

You certainly do not want to bring infestation into your plant on incoming products. It is important to carry out checks on susceptible raw materials on arrival. The Auditor will look for a system of checking and will want to see clear records of such checks. This is likely to be at the 'Goods In' stage and should include bulk materials as well as bagged or boxed ingredients.

Example

A wine manufacturer was found to have no system of checking or recording for infestation of incoming grapes.
A nonconformity was given.

Part Two

Clause 4.13.1

Essentially, you have two options for satisfying this requirement. You must either enlist a contractor for pest control (and most companies do) or you may use your own staff. In both cases, competency must be demonstrated. The Auditor will ask for evidence of this, usually in the form of certificates of qualification.

Example

A cheese factory used a contractor but the pest controller's qualification certificate had lapsed and staff training (rodents only) was unclear on competence.
   A nonconformity was given.

You must also arrive at inspection frequencies by risk assessment and document this. If you are using a contractor, you should ask them to provide evidence that they have done this, so that it is available for the Auditor.

Whether you use a contractor or not, you must ensure that the whole site is properly covered.

Example

Not all areas of the site had been included in the pest control programme, notably no internal rodent bait boxes located in the ex-freezer store, which had become an ambient store.
   A nonconformity was given.

The final part of the clause requires that where a contractor is used that the contract is clear and meets the needs of the site. The Auditor will ask to see the contract (so ensure that it is available) and will look to see that all likely pests have been covered for your product, that inspections are sufficiently frequent and that the contract covers the whole site.

Examples

At a produce manufacturer, it was found that the pest control contract did not currently include the new canteen building.

The pest control contract for a milk dairy was found to be deficient in that roof voids did not appear to be monitored under the contract.

At a manufacturer of cooked meat, there were no externally sited rodent bait boxes at Unit 2 (the unit was sited directly opposite a long-disused building).

Another produce manufacturer was found to have a pest control contract that did not clearly indicate or confirm the frequency of catch tray analysis.

Nonconformities were given in each case.

Clause 4.13.2

This is a new clause for Issue 6 and is a specific set of requirements for those companies who carry out their own pest control. Feel free to move on if you use a contractor. If you carry out your own pest control, the five bullet points are all worth looking at in detail.

Training of the pest control staff is very important and the auditor will need to see some evidence of this. In some countries, this can be provided by pest control associations, in others, perhaps by agronomists. It is important because pest control requires specialist knowledge of the behaviour and life cycles of pests and of the chemicals to be used especially with regards to the limitations of use in a food factory.

Examples

The on-site Quality Manager conducts five pest control inspections per year, as part of the current pest control contract. However, there was no evidence of training for the Quality Manager.

At a pitta bread factory, the Auditor found that an untrained site employee was responsible for the application of insecticides.

Nonconformities were given.

Resources are important in this situation in terms of finance and manpower. Doing your own pest control may require sufficient staff to have deputies in case of incidents during absence.

You will also need to demonstrate that you also have access to specialist knowledge when required. For example, you may need to identify a pest or you may have a recurring issue of infestation that you have been unable to resolve yourselves. This would be of some concern to the Auditor (and should be to you). Aside from pest contractors, you might use certain university laboratories for example.

Pesticides are controlled by law in most countries and certainly their proximity to foodstuffs should be controlled. If you are doing your own pest control, you will need to demonstrate understanding of this to the Auditor. If you store pest control products including pesticides on site this shall be in a locked facility.

Note: If you carry out your own pest control Clause 4.13.8 might be of special significance.

Part Two

Clause 4.13.3

This requirement is for documented procedures and general documentation concerning your pest control system. No doubt if you use a contractor their documents will help in meeting some of these requirements but not all. There are four features that you must have as a minimum.

The first of which is a very straightforward requirement for all pest control measures to be shown on a site plan. Clearly, the Auditor will need to see this so make sure that it is available. You should also make sure that it is up to date, signed and shows all areas. It must show all pest control measures, for example rodent baits, EFKs, moth traps and insect monitors. Where items are numbered, they have to be identified on the plan.

Although it should be straightforward, nonconformities often arise with the site plan.

Examples

At a manufacturer of sliced, cooked meats, the Auditor noted that the pest control plan was not updated to show the current provision.

A rice manufacturer had a pest control plan but there was no plan for the external store.

At a tomato cannery, the Auditor found that locations of the EFKs were not identified on the pest control plan.

A Chinese food manufacturer, there was no plan at all of the pest control measures.

Nonconformities were given in all cases.

So, you need to make sure that the plan is comprehensive and up to date.

Aside from identifying all the baits and monitoring devices you must also define responsibilities for the management of pest control on site. The Auditor will ask who is responsible for pest control. This person is likely to liaise with any contractor and should also be the person responsible on a day-to-day basis. Evidence for this should be available and may be included in your support documents showing all responsibilities (see Clause 3.3.2).

Whether or not you use a contractor, your procedures must also include details of the pest control products and instructions for their use. Contractor's Manuals often come with this kind of data but you must check that it is all present.

Example

At a produce packer, it was found that there were no records of the details of the toxic bait used for rodents.
A nonconformity was given.

Your records must also include any observed pest activity and details of any treatments carried out (see also Clause 4.13.7).

Clause 4.13.4

This clause specifies some important requirements concerning baits. In addition to being appropriately located, they must be robust, tamper-proof and secured in place. The Auditor will check on this during the site tour. It is essential that you have specified these features with any contractor.

Examples

Rodent baits not secured in place and not all shown on map.

At a produce packer, it was noted that rodent baits were not secured. There were two missing in the packaging store, possible as a consequence.

A nonconformity was given in each case.

Toxic baits should be used with great care and on a restricted basis. Thus, regarding rodent baits within open product areas, in a new requirement for Issue 6 you shall use non-toxic indicator baits rather than toxic baits unless you happen to be treating an active infestation. In such cases, you must take care to ensure that there is no risk to product.

Clause 4.13.5

This is a simple clause, the first part of which dates back to the first issue of the BRC Standard. This is the requirement for EFKs or pheromone traps to be correctly sited and operational.

Example

At a dry ingredients manufacturer, the Auditor was concerned to find no pheromone moth traps sited in the new final product warehouse.
   Because these were flour-based products that could be susceptible to flour moth, a nonconformity was given.

The correct siting of EFKs and moth traps is essential not just for their effectiveness but also because in the wrong place they can be a hazard themselves to food.

Part Two

An EFK sited over exposed food might cause 'fried' insects to bounce off into the product.

Examples

At a bean cannery, it was noted that an EFK was located directly above the bean soaking tanks.

At a fresh fish processor, two EFKs were sited directly above the filleting lines.

A nonconformity was given in each case.

In cases such as these, if it is not possible to re-site the equipment, you must use alternative systems.

Because they give out light, EFKs are very obvious in a food factory so the Auditor will easily pick them out and see if they are inappropriately sited and will readily see any that are not working. It is advisable to check your EFKs regularly to make sure that they are operating. Also, because their efficiency diminishes, it is considered standard practice to have a change of lamps annually, preferably before the worst months for insects.

Example

A sandwich manufacturer had a schedule for changing the EFK lamps but the Auditor found that the EFK bulb change had not taken place as scheduled.
A nonconformity was given.

Regarding flying insects, EFKs work well in most situations and they can be useful in storage areas as well as production areas. However, an often neglected aspect is light fittings especially those with heat vents which allow insects in. An Auditor can often tell a lot about the level of insect infestation in a site by looking at the lights because, while EFKs are usually cleaned out regularly, light fittings can be ignored, particularly in storage areas.

Example

In the main dry goods store of a bakery, the Auditor found in excess of 50 dead wasps in a light fitting.
A nonconformity was given.

Do not ignore light fittings, keep them clean: they can be a good indicator to you of levels of insect infestation.

Clause 4.13.6

This clause requires that you take immediate action in the event of infestation. The Auditor will look at your procedures to see that they encompass this and will also check records for actual infestation to see what action was taken and how quickly. You will be expected to have eliminated the infestation and ensured that any affected product was suitably isolated and evaluated.

The action that you take will obviously depend on the nature of the infestation. In the case of insects, the use of insecticides may be required to knock down the population but you must also consider eliminating the cause of the outbreak and thus preventing the problem. Thus, proofing or housekeeping might need to be reviewed.

Affected products should be treated as nonconforming and subject to the procedure for nonconforming product (see Clause 3.8). Of course, the need for isolation is acute if there is active infestation in a product.

Clause 4.13.7

One of the more common failings in pest control systems is the failure to keep good records of inspections and of any corrective action necessary. The Auditor will want to have a good look at pest control records, probably for the year leading up to the audit. If you have changed contractor in that time, ensure that you have all your records available. Some companies now offer an internet-based reporting system, so you may need to have access ready to show the Auditor.

The Auditor will want to see records of each inspection with full reporting of all areas inspected. Any issues noted such as evidence of pests, proofing issues, poor hygiene, which could lead to pests and so forth shall be clearly recorded along with any corrective action responsibilities. Corrective actions should be signed off in good time.

Part Two

Examples

At an ice cream factory, the Auditor noted a hole in a door that required proofing but found that the pest controller had not identified it. Also, corrective actions were not signed off.

A poultry manufacturer was found to have no corrective actions recorded against the Pest Controller's recommendations to remove crates/surplus equipment alongside the external wall.

> At a cheese packer, the Auditor found no record maintained of pheromone moth trap activity.
>
> A food ingredient manufacturer was found to have pest control records that were not always signed off by the site for completion of corrective actions.
>
> At a processed cheese supplier, the Auditor found that a lack of access by the pest controller to two bait points has been reported to site on the last four occasions but no remedial action to allow access had been taken.
>
> Nonconformities were given in each case.

Clause 4.13.8

This is a new clause for Issue 6 of the Standard and requires that you have an in-depth survey carried out by a pest control expert, that is someone with an in-depth knowledge of pest control. The frequency of this shall be risk assessed but the clause suggests every 3 months. This is typical of the type of service offered by many pest control contractors, so for many companies it should not be a problem. Those carrying out their own pest control will need to consider this. Unless you can demonstrate to the Auditor that you have sufficient expertise in-house, you will need to seek help outside the Company.

Examples

> The Field Biologist's report for the visit scheduled for that year was not in evidence.
>
> A jam factory was found to have insufficient pest control site inspections; notably no Field Biologist/Technical pest control site inspections, and only four routine Technician inspections per year.
>
> A nonconformity was given in each case.

Clause 4.13.9

This is a requirement for the trend analysis of pest control results. In most cases, pest control companies can produce statistics as part of the service and if so the Auditor will expect to see this in the contract and see the figures in the reports. Of course, it is also something that you can do yourself.

Even if the pest controller is doing, say, EFK tray counts for you, you must also show evidence that you have taken account of and reviewed the data yourself.

Examples

At a pasta factory, the Auditor found no evidence of data analysis of pest control measures.

At a small bread bakery, there were no counts, records, or trending of pheromone moth traps undertaken.

At a wine bottling plant, there was no trending of rodent activity in pest control records.

At a sandwich maker, the results of pest control inspections, with the exception of EFK catch-trays, were not assessed and analysed for trends.

All were given nonconformities.

Such analysis must be carried out at a regular frequency but at least annually and when an infestation has occurred. Note that this is not only about EFKs but also all pest control data.

Clause 4.14: Storage Facilities

For Issue 6, we finally have a section devoted solely to storage and not before time. The SOI presents the objective that all storage facilities shall be suitable for purpose.

Key points in Clause 4.14
Procedures
Temperature monitoring
Controlled atmosphere storage
Handling storage outside
Ensuring stock rotation

Clause 4.14.1

This is a requirement for documented procedures to cover all aspects of storage. To meet the SOI, you must consider ingredients and in process products.

Example

At a bread bakery, the Auditor found that a number of white shorting (Fat) blocks, which were intended for use on the next day were left uncovered on the puff pastry conveyer belt.
A nonconformity was raised.

There are four specific aspects to consider, all of which should be standard procedures in a well-run storage facility. The procedures must be based on risk assessment: not all the possible inclusions will apply to every site. The four aspects are as follows:

(1) Where temperature control is essential, you must have procedures for maintaining temperature during any moments or transfer between different areas. (General temperature control in storage is referred to in Clause 4.14.2). Thus, for example, chilled or frozen products should not be left for too long a time in ambient areas.

(2) An important requirement concerns segregation of incompatible materials, in particular for volatile products and those that may pick up taints of such. Flavourings, high-fat foods, fats and oils are noted for being especially susceptible to taints and there are other examples such as dried pulses. Note that the source of taint could also include non-foods such as cleaning chemicals. The Auditor will be aware of the nature of your products and certainly if there are any susceptible ingredients on site will be looking for adequate segregation during the site tour.

(3) In storage, you must ensure that materials are off the floor and away from walls. There are no minimum distances indicated but common sense will apply. You should facilitate both routine cleaning and inspection for pest control. Obviously, this is something that Auditors will be looking for during the site tour.

Examples

A manufacturer of cakes, doughnuts and croissants was noted to have raw material stored directly on the floor in the production area (lard in cartons).

At a jam factory, ground level palletised materials in Warehouse were stored directly against the walls, with no access for pest control or cleaning.

A nonconformity was given in both cases.

(4) Finally, your procedures might need to include specific handling or stacking requirements such as always using racking.

Clause 4.14.2

This clause is about aspects of temperature control in storage. Firstly, it is concerned with the capability of the facility. Be mindful that this can include product that requires to be kept warm or tempered as well as chilled or frozen.

Here, you must be able to demonstrate that storage facilities are capable of maintaining the required product temperatures and are operated to do so. This requires equipment of the right specification and the Auditor will ask how you have ensured this. You need to consider insulation, airflow and many other factors.

Secondly, the clause is about the maintenance of product temperature which, of course, the Auditor will look at closely during the site tour and when looking at records. The requirement allows for both automatic temperature recording equipment and manual systems but basically you must have a recording system in place. If you have continuous monitoring of storage temperatures you must also have am alarm system. If not, you must have a system of manual checks in place. How often should you take the checks? The position of the Standard on this is that the frequency of checks is governed by how long the product can safely hold its temperature if the system fails. That will be up to you to establish and justify; however, the Standard also indicates that 4 hours should be typical. Whatever you decide, you must then ensure that these checks are carried out around the clock, 7 days per week. Weekends should be covered.

Examples

A manufacturer of natural set yoghurt did not work at weekends. The Auditor found that they had a finished product chilled storage area, which was not monitored during the weekend, nor was their any alarm system in place in case it failed.

In a meat pie factory, temperature-controlled areas were being monitored manually, but there were no records for Saturdays or Sundays.

A nonconformity was given in each case.

Part Two

Clause 4.14.3

This is a new clause for Issue 6 concerning controlled atmosphere storage. Thus, it could include the use of gases such as in the storage of some fruits and vegetables, or a certain humidity. The Auditor will need to see that you have specified the systems and that they are controlled. You must also keep records of the storage conditions.

Clause 4.14.4

Clearly, any food materials that are stored outside must be protected. An example sometimes seen is ingredients in large metal drums such as tomato puree. In such cases where the container is exposed to the weather, the Auditor will be looking for safeguards in your procedures to ensure that the contents are not contaminated when you come to use them. A container that is covered in bird droppings will obviously need some attention! Any packaging that is stored outside such as cans or glass jars must be similarly protected.

The clause also applies to equipment and pallets. It is not uncommon to see wooden pallets stored outside. Precautions must be in place to prevent any contamination of product when such items are brought into the factory.

> ### Examples
>
> At a produce packer, some product bins ('cagettes') were held outside after cleaning with no protection.
>
> At a manufacturer of jams, Externally stored ingredients were not adequately protected; notably unlocked bulk apple concentrate silo intake pipe, unlocked box containers holding concentrate drums and packaging.
>
> A nonconformity was given in each case.

Clause 4.14.5

This requirement needs no explanation. Most packaged items received will be accompanied by documentation and most will be labelled with a durability date. Where this is not the case, you may have to create documents or label products yourself to ensure correct stock rotation. The Auditor will look at products in storage and products that do not bear any date or rotation number will be called into question and may lead to a nonconformity.

> ### Examples
>
> At a manufacturer of a range of products including soft drinks and chocolate confectionery, the Auditor found that a sugar pallet code number was not available as it had been discarded: there was no procedure in place to retain ticket/lot number.
>
> In the raw material warehouse of a dry ingredients manufacturer, the Auditor found pallet labels for paprika and turmeric, which were incorrect.
>
> In the small freezer at a bakery, there were a number of boxes of out of date (13 March) margarine and the frozen cream dated 17–21 July did not match the dates logged on the freezer log of deliveries.
>
> Nonconformities were given in all cases.

Regarding raw materials, their rotation is sometimes referred as 'first in – first out' or 'FIFO'; however, it should be remembered that it is possible for a supplier to deliver goods that are older than an earlier delivery, so the use of products in the correct order should refer more to their shelf life than when they were actually received.

The Auditor will want to see a positive system in place for this, an occasional stock check alone may not meet the requirement. Ideally, computerised stock control systems can be employed which drive the use of the oldest material first; however, this can still be adequately done with paper-based systems. Even so, it is still possible for human error to affect this, so stock checks can still play a significant part in a system.

Examples

At an Indian meal manufacturer, the Auditor found items of mixed durability dates in storage such as colouring material. There appeared to be every chance that this would be used out of chronological order.

At a bakery, out-of-date soya flour and vanilla concentrate was noted.

At a manufacturer of soft drinks, the Auditor found raw material (egg white) in the refrigerator that had passed the 'use by' date.

A nonconformity was given in each case.

Note that this issue seems to arise more often with small ingredients, so be careful about your rotation of them.

Clause 4.15: Dispatch and Transport

This SOI covers all aspects of loading and transport. As always, the SOI has the safety and quality of product as the objective and that you have procedures in place to ensure this. In essence, transport is an extension of your factory and should be treated as such.

Key points in Clause 4.15
Documented procedures
Load inspections
Ensuring traceability right to the end
Vehicle inspection
Temperature maintenance
General maintenance
Handling mixed loads
Breakdown procedures
Security of unattended vehicles
Use of contractors

Clause 4.15.1

This is a requirement for procedures (documented, of course) to cover all aspects of transport and the clause lists four possible aspects to consider:

(1) Where temperature control is essential, consider measures for controlling temperature in loading dock areas.
(2) Another point concerns unloading or loading in covered bays. This clause concerns materials that are susceptible to weather damage and the need to protect them. The term 'covered bays' can be taken to include open areas with a roof or docking bays, which are sealed. The Auditor is likely to take a pragmatic view on what types of product are susceptible. Certainly, exposed foods would be at risk as

Part Two

would products, which are in bags or cases which would take up moisture in the rain or snow. Products, which are sufficiently sealed to protect them from the elements are not at risk.

(3) This point concerns the securing of loads to prevent movement during transport. This may require vehicles to have load supports, lashing points and so on.

(4) Where appropriate, all loads to be inspected prior to dispatch. This might not be appropriate for material such as bulk powders or liquids, which cannot be inspected but otherwise this should be in your procedures and the Auditor is likely to ask for evidence of your routine inspections in the form of records.

Clause 4.15.2

Traceability is discussed in detail in Clause 3.9. This requirement is intended to ensure that traceability is continuous through dispatch and transportation. In particular, the Auditor will want to see evidence of traceability in your dispatch records, so you must have these available. In some cases, you may also need to ensure the correct documentation at the delivery point.

This may be of particular importance where a customer has numerous delivery points and you may need to trace where a batch of product has gone in order to carry out an up-lift.

Example

A UK retailer with several depots was supplied with cooked meat pies by a UK supplier. On one occasion, they mis-coded a product, giving 31 June as the Use By date. As this date does not exist, all such labelled product had to be withdrawn.

'No problem' said the supplier, the product only went to two of your depots. So all affected product was withdrawn from those two. However, some days later, other depots were finding the problem in their area.

Clearly, the supplier did not have a foolproof system of forward tracing batches to all the depots of this particular customer and were asked to make significant changes to their systems to ensure that they could do this in future.

Clause 4.15.3

Considering transport, vehicles must be hygienic and designated for food use. Clearly, no volatile material should be carried along with foods likely to be tainted. However, in this case, we must also consider the previous history of a vehicle's use. Certain materials are difficult to remove completely. Furthermore, vehicles or shipping containers, which have been used for unsuitable non-foods, for example chemicals, must be avoided. The spoiling of dried pulses with phenolic taint from ships' holds is a classic example.

Therefore, vehicles and containers must be inspected *before* loading and you will need to have records of this, which are available for the Auditor. The inspections and records must demonstrate that you have observed the four bullet points of the clause, that the vehicles/containers are clean, free from odour or taint, well maintained and equipped with any necessary temperature control.

Example

At a bread bakery, the Auditor was concerned about the condition of light fittings within the dispatch vehicles, yet there was no reference to internal light integrity within dispatch vehicle pre-load checks.
    A nonconformity was raised.

Clause 4.15.4

The first part of this requirement is for transport to be capable of meeting the product temperature requirements. This must take into account any variation in the loads from minimum to maximum.

Secondly, you must show evidence either of continual temperature measurement such as the use of data loggers, or some other means of validating the temperature in transit. Data loggers now come in very convenient forms and can be used to monitor temperature over a range of periods. On the other hand, if you are using manual recording, this must be done at predetermined intervals, so in this case the Auditor will want to see a work instruction or procedure to this effect. The Auditor will want to see records to demonstrate this.

Part Two

Examples

At a manufacturer of potato-based products, there were no records of data-logger temperatures from dispatch vehicles available for the Auditor to examine.

A Chinese food manufacturer was found not only to have no maintenance records of the company vehicle, including the freezer unit but there was also no regular temperature monitoring during deliveries.

Not all site-owned/operated chilled dispatch vehicles maintained a temperature record throughout the journey to customers.

At a fresh meat manufacturer, it was found that the data-logging device for monitoring vehicle temperature was not working.

A cheese and dairy products manufacturer were found to have no temperature monitoring or records maintained for their new 7.5-tonne chilled despatch vehicle.

Nonconformities were given in each case.

Clause 4.15.5

You must have maintenance systems and cleaning procedures for all vehicles and ancillary equipment such as delivery hoses. The Auditor will ask to see these procedures. The Auditor will be looking for procedures that ensure vehicles are in good hygienic condition before they are loaded.

> ### Examples
>
> At a sugar confectionery site, the Auditor found no documented procedure to ensure vehicle suitability and vehicle cleanliness.
>
> At a pasta factory, the Auditor found that there was no evidence of verification of vehicle cleaning for durum wheat and finished product.
>
> Nonconformities resulted in each case.

Thus, the Auditor will also inspect vehicles and where applicable items such as inlet hoses during the site tour. Typically, inlets and hoses should be kept capped when not in use and inlets are usually expected to be locked also.

Clause 4.15.6

This clause concerns your procedures for product transport and has three points, which we shall consider:

(1) You must have documented procedures on any restrictions on mixed loads. By this is meant any mixing of products, which might be risky such as a mix of chilled and ambient products. Alternatively, you might need to show how you avoid cross contamination or taint in mixed loads.

(2) This point concerns security during transport. This subject can be difficult for companies and may require new thinking. Of most concern are fully loaded vehicles, which are left unattended. You need to have clear instructions in place for this scenario.

(3) Finally, there is a requirement for breakdown procedures for vehicles and where appropriate for refrigeration units on vehicles. The Auditor will want to see what measures you have in place for either occurrence and, of course, this must apply to any contractors as well. Because we are dealing with matters most likely occurring off site, the Auditor will want to see documented procedures for this aspect.

Examples

At a manufacturer of pitta breads, the Auditor found no documented dispatch vehicle breakdown procedure available.

A poultry manufacturer did not have a dispatch vehicle breakdown procedure not formalised with the current haulier.

Nonconformities were given in each case.

The procedures must include instructions relating to assessment of the safety of products during such a breakdown and you must also maintain records of such incidents.

Example

A sausage and meat manufacturer was found to have had vehicle and vehicle fridge breakdowns but they were not formally recorded and corrective actions were not closed out.
     A nonconformity was given.

Clause 4.15.7

Where you use a contractor for transport, you must have a contract that defines *all* the requirements of Section 4.15 of this Standard. Also, you must either verify this yourself or the contractor shall be certified to the Global Standard for Storage and Distribution or other similar internationally recognised standard (but we won't mention them).

Examples

A manufacturer of sachet products for catering was found to have no contract available to show that the contract hauliers met the Global Standard requirements, that is that they be clean, suitable, in good repair, free from taint and so on.

Similarly, a sugar confectionery factory was found to have no contracts in place for the third-party contract companies that they were using.

At a stewed fruit dessert manufacturer, the Auditor found that there was no comprehensive contract available for chilled final product dispatch hauliers, detailing transport temperature, vehicle maintenance, vehicle hygiene and so forth.

Part Two

A hot dog manufacturer was found to have no comprehensive contract for frozen dispatch transport companies, detailing all requirements of Global Standard Clause 4.15.

At a cake and muffin factory, there was no reference in the third-party final product haulier contract to ensure that mixed loads of final products would not risk contamination.

Nonconformities were given in each case.

Summary

This section is the largest in the Standard and much of it will be audited during the site tour. However, remember that there are many requirements for procedures and records here. No part of a well-run food safety system can exist without the paperwork. This section might appear to be about the physical facilities but it is also about staff, their facilities, their well being and their training. Interviewing staff will form a major part of the auditing of the section.

Quiz No. 9

(1) Where can you use toxic rodent baits?
(2) How often shall you review security arrangements?
(3) How often must you test water quality?
(4) What does a Gauss unit measure?
(5) What four process parameters must be defined and validated for an effective CIP system?
(6) How often should you make temperature checks on a storage system with no automatic recording equipment or alarm?
(7) When do you need to have covered loading bays?
(8) What sort of products may not require metal detection?
(9) Where must you provide hands-free taps?
(10) You must provide a plan of the drainage systems for all factories: true or false?
(11) Where might you use non-potable water?
(12) Where may you not offer visitors shoe covers?

15　Clause 5: Product Control

Clause 5 covers some apparently diverse subjects as product development, handling of allergens, product analysis and others. The common theme is that they are all elements of control within your site. Although the section is largely colour-coded green, you can expect much of it to be covered during the site tour as well as the document review.

There is one Fundamental Requirement in this section, so I am changing the running order and will deal with that first. Note that this Fundamental Requirement is the only one in the Standard to have changed its title for Issue 6 as it now is solely concerned with allergens (identity preserved products are considered in a separate clause (Clause 5.3)).

Clause 5.2: Management of Allergens (Fundamental)

> **Key points in Clause 5.2**
> Managing all allergens
> Ensuring all sources are considered
> Make a list
> Where will cross-contamination occur?
> Preventing cross-contamination
> Rework
> 'May contain?'
> Ensuring proper allergen cleaning
> Training
> Labelling

The BRC Global Standard for Food Safety: A Guide to a Successful Audit, Second Edition. Ron Kill
© 2012 John Wiley & Sons, Ltd. Published 2012 by John Wiley & Sons, Ltd.

Part Two

This is an important SOI; it is a Fundamental Requirement because the issue of the handling of allergens is so crucial to food safety. Obviously, to those who suffer a food allergy it has always been important; however, it is only in the last decade that the consciousness of the food industry has been raised to this issue so much. Although there are some doubts expressed by some as to the validity of the numbers, there is undoubtedly a perception that numbers of cases of food allergies among children are on the rise ('Why are food allergies on the rise': CNN news item August 2010; 'Doubts raised over child food allergy rise' BBC news item August 2010).

It has become increasingly important in Europe since the publication of the legislation based on directive 2003/89/EC. Retailers are especially concerned about this matter. In the past 4–5 years, the FSA in the United Kingdom has issued many food alerts concerning products that have labels that do not name all allergens present. The SOI objectifies a management system for allergenic materials handled on site to minimise the risk of contamination.

In the past, some companies might have been guilty of doing little or nothing to control segregation of allergens in the sites, relying heavily on disclaimer labelling such as 'may contain...'. Now, however, it is emphasised that companies must make all efforts possible to segregate materials and only then, if there remain possibilities of cross-contamination shall a company use such disclaimers (see Clause 5.2.6). This is an important change in emphasis for Issue 6 and Auditors will look closely at this aspect. The SOI must be satisfied and in order to meet this you have to make an effort! The SOI also requires that you meet the legal requirements for the labelling of allergens, thus for products sold in the EU, for example, all the major serious allergens listed in the Glossary to the Standard must be labelled if they are in the recipe. They are as follows:

Cereals containing gluten
Crustaceans
Fish
Eggs
Peanuts
Soybeans
Milk
Nuts (namely, almond, hazelnut, walnut, cashew, pecan nut, Brazil nut, pistachio nut, macadamia nut and Queensland nut)
Celery
Mustard
Sesame seeds
Sulphur dioxide and sulphites at levels above 10 mg/kg or 10 mg/l, expressed as SO_2
Molluscs
Lupin

The Auditor will want evidence of how you control such materials in your site and will check such practices during the site tour.

Example

A manufacturer of dried potato products caused the Auditor concern because following allergenic raw material usage, for example egg powder, no records were maintained of the prevention of cross-contamination during food contact surface clean-downs in the mixing or filling departments.
    A nonconformity was given.

Note: Remember, when considering allergens, consider any country where the product will be sold because some have different lists of allergens.

Clause 5.2.1

This requirement can be summarised as follows. You must carry out an assessment of raw materials. This is to establish the risks caused by the presence of allergens and any possible cross-contamination. You should also be aware of any possible allergens present in additives, for example as carriers of flavours.
    The first thing to do is to assess and approve raw material specifications with this aspect in mind. They may be allergens themselves or they may contain them in small amounts. Raw material supplier sites may have difficulty in absolute segregation of such raw materials and risk assessment technique will assist you in determining proper controls. You will need information from suppliers to assess this. This can be established by supplier questionnaires that specifically focus on this matter. Remembering Clause 3.5.1.2, you may also look at this aspect during a supplier audit.

Part Two

Examples

A beer brewery was found to have not carried out an allergen risk assessment for all raw material ingredients. Clearly, for beer making, there are potentially allergenic materials such as barley present.

A manufacturer of chilled and frozen ready meals was found to have excluded celery (which was present on site) from their assessment of allergens.

A processor of fruits and vegetables was found to have no confirmation of sulphite levels in dipped potatoes.

Nonconformities were given in each case.

Remember to include aspects of raw materials possibly being contaminated by your supplier.
    Note: This is not only about nuts but about all allergens.

> ## Examples
>
> A manufacturer of cheese, dairy products and certain ambient products was found not to have considered reference to presence of allergens in their supplier approval questionnaire system.
>
> A chocolate manufacturer was found to have a supplier approval questionnaire that did not include the requirement for suppliers to provide details of all potential allergens; only nuts were considered.
>
> At a rice mill it was found that risk assessment has not been undertaken on the loading of finished rice into bulk tankers that may have previously transported wheat.
>
> Nonconformities were given in each case.

Clause 5.2.2

What you must do to satisfy this requirement is consider all your raw materials intermediate products and finished products and list those allergens in them. The point of this requirement is that any segregation systems you have, might need to reflect all those stages.

The Auditor will want to see this list, so have it available.

Clause 5.2.3

As we looked at previously, you must do everything possible to avoid cross-contamination of allergens at all stage. So, in this clause, is set out the means for this. You must carry out a documented risk assessment to identify routes of contamination. You must also create procedures (documented, of course) for handling all stages of product to avoid cross-contamination. In particular, the clause specifies four aspects that you must include in your risk assessment, which include consideration of the nature of the material. Airborne dust might be a problem, for example (see also Clause 5.2.4). Remember to include all the allergens on site.

> ## Example
>
> A bakery's allergen risk assessment had not considered celery processing (a relatively new process to the vegetable preparation area) or dairy as a bakery ingredient.
>   A nonconformity was given.

The way of thinking here is somewhat HACCP-like, because the clause asks you to think about the nature of the material that might indicate the likelihood of a hazard, then, looking at the process flow, the potential points of cross-contamination. Once you have established them, you will consider the risk of cross-contamination at these points and any controls you can put in place to reduce or eliminate them. The language is indeed very HACCP-like and the HACCP team might be an appropriate group to look at this clause.

The Auditor will want to see your documentation and, despite the clause being green colour-coded, will have these control points in mind during the site tour.

Clause 5.2.4

You have done your risk assessment and identified any necessary controls to prevent cross-contamination. Now, you must write procedures to ensure this. The clause specifies seven aspects that you must include as appropriate. Correct storage is extremely important of course.

Example

A bakery had inadequate allergenic ingredient storage; notably nuts stored (open bag within an open box) on top of other ingredients on a pallet in the bakery, and unpacked shell eggs beside bakery yeast.
   A nonconformity was given.

One of the points is the use of identified, dedicated equipment and food brought on to the site by staff.

Examples

At a bread bakery it was found that trays with allergens (colour-coded yellow) were not stored on the bottom of the racks to minimise contamination.

While the bakery met part of the requirement in that they were using identified, dedicated equipment for allergens (the yellow trays), a nonconformity was given for failing to segregate them sufficiently.

More clearly, at a biscuit bakery, there were no dedicated ingredient scoops for individual allergenic ingredient handling. Equally worrying at the same site, liquid egg was inappropriately decanted into a 'washed' ingredient container that formally contained another allergenic sub-ingredient (soya).

Nonconformities were given in each case.

Another concern is the food that staff bring on to the premises. This may be particularly appropriate if you are making claims about your products. For example, some

Part Two

sites now wish to make the claim that they are nut free and have consequently extended this to what foods staff are allowed to bring on to site for their own consumption.

The scheduling of production is a very important aspect of control. Clearly, if you are using common production equipment, it makes sense to run allergen-free products first. To minimise changes during the day is also good practice here.

The other four issues are (1) physical or time segregation, (2) the use of separate protective clothing, (3) systems to restrict movement of airborne dust and (4) waste handling. The last point is something that could be related to your site plan, which must include the movement of waste (see Clause 4.3.2). The Auditor will be looking at where materials are stored, how equipment is used, the running order of production, cleaning down between different materials, staff handling procedures and so on. Any documented procedures that you have to support this should be available for examination.

Example

A company that manufactured ice cream and dairy products but that also supplied other products bought in did not have comprehensive systems. While all raw materials requiring specific handling (nuts) were controlled, bought in nut products for distribution were not segregated from other materials.
    A nonconformity was given.

This clause is colour-coded peach but the Auditor will look at the procedural aspect during the document review.

Clause 5.2.5

This is a very specific requirement relating to rework. Naturally, if you do not create and process rework, you can ignore this requirement. However, there are many industries that use an element of rework and it is good general practice to ensure that you have procedures in place to ensure product safety, legality and quality. Where a rework cycle is potentially continuous, the cycle must be broken at certain intervals. Here, your requirement is that you must have procedures to ensure allergens are not an issue and that rework containing allergens does not find its way into material that is free of those allergens.

This is a procedural matter but again the Auditor will also take in actual practices during the site tour.

Clause 5.2.6

This is a new clause for Issue 6 concerning the type of disclaimer labelling that is common in the United Kingdom. As we have stated previously, your whole emphasis must be on prevention. However, when (and only when) you have taken all possible steps and there still exists the possibility of cross-contamination then you must include a warning on the label to indicate to the consumer that the product might contain certain allergens. The clause requires that you meet national or international guidelines on this matter. Currently, the United Kingdom, having flirted with such phrases as 'made in a factory that also handles X' and such, like the recommendation from the FSA, at the time of writing, is for 'may contain X' (grammatically poor but at least it is succinct) or 'not suitable for someone with X allergy'. You must check for your own country or for any region you are exporting to.

Making no effort to prevent cross-contamination and relying on a 'may contain' statement will not meet the requirements.

Clause 5.2.7

This is another clause concerning claims on labels or other publicity material (see also Clause 5.1.6). In this case, it concerns claims about suitability for allergy or sensitivity sufferers. Thus, 'gluten free' or 'suitable for coeliac sufferers' would be a positive claim. This is not about warnings or the 'may contain' statements mentioned previously.

The Auditor will determine if you are making such claims and if so will want to see that you have validated that your process will meet them. You should consider not only your own raw materials and production environment but also those of your raw material suppliers. Thus, if you claim a product to be nut free, you may need to consider the source and transport of your raw materials to ensure that there has been no cross-contamination before it gets to you.

You must document this validation, so have this evidence available for the audit.

Clause 5.2.8

Clearly, all precautions that you make to ensure that there is no cross-contamination that breaches your allergen controls will be undone if your cleaning systems do not facilitate the continued segregation and/or protection from unwanted allergens. To satisfy this requirement, you must have documented cleaning procedures that, based on risk assessment, reduce the potential for cross-contamination, or remove it altogether.

The Auditor will consider these aspects when looking at your cleaning procedures (see Clause 4.11.1) therefore, you must make clear in them any special methods you use to tackle cross-contamination with allergens.

Examples

A manufacturer of muesli bars had some products that did not contain nuts. However, the Auditor found that there was no comprehensive process line clean down of the process line between nut and non-nut production runs.

A manufacturer of pitta bread and sesame seed buns had a clear obligation to ensure that sesame seeds did not contaminate product that did not contain them. They had gone some way with this but the Auditor found that the allergen-control procedure was not finalised; notably the post-sesame seed production run clean down was not comprehensive.

Nonconformities were given in each case.

The Auditor will also want to see records that demonstrate that you have validated cleaning to ensure no cross-contamination. This will mean some positive reporting of this aspect. The validation of a cleaning method is something that you need not necessarily perform every time that you clean. When we talk of validation we mean ensuring that the method is correct. It is possible that you might need to do this only once in a while. Depending on the situation, this validation may require some testing and there are testing kits available for this. You may choose to validate cleaning procedures by using the test kits on cleaned surfaces. They are not yet available for all allergens but for a good number. Another way might be to send finished product made just after cleaning away for analysis.

Example

A bakery that also handled produce was unable to show any validation evidence of the allergen-cleaning procedures, notably, celery dicer, bakery mixing bowls and other food contact surfaces following production of bakery products containing nuts, sesame seeds, dairy products, sulphite in potatoes (post-Dry Wite washing).
A nonconformity was given.

However, you also need to verify cleaning. Verifying is different and this should be done each time you clean.

Clause 5.2.9

This requirement concerns training. The Auditor will ask to see records that demonstrate that the personnel concerned have received appropriate training in the matter of allergen awareness. Note that this includes temporary staff and contractors, so even non-employees coming on to site must receive some training.

Example

A confectionery supplier handled some allergens on site. The Auditor considered that relevant staff had not been adequately trained in the handling of allergens.
A nonconformity was given.

Note: This clause is dual colour-coded, so the Auditor may interview personnel as well as look at records.

Clause 5.2.10

This clause concerns labelling. At the beginning of this section on Clause 5.2 it was mentioned that there have been many instances of wrongly labelled products leading to product recalls because the wrong information about allergens had appeared on products as a result. This clause is written to address this. It requires documented checks at start up, product or packaging changes and to prevent mislabelling. Clause 6.1.7 states much the same thing for all product handling anyway, so these are processes that you will need to have in place even if you do not handle allergens. As we have stated elsewhere, the Auditor will ask to observe such changeovers during the site tour.

Clause 5.1: Product Design/Development

Now, we return to the start of this section. This is a straightforward clause but there are two aspects. Firstly, the SOI concerns product design and development procedures, required to ensure that safe and legal product can be manufactured. It also refers to any changes to product packaging or processes, which must also be considered in the clauses. In the past, some considered that this section only referred to NPD and for many companies and indeed Auditors who considered that they did not do this, it was left somewhat blank. The change in this SOI for Issue 6 means that Auditors will be asking for at least some of these clauses to be met because it now includes any changes to product or packing.

Part Two

> **Key points in Clause 5.1**
> Managing all change
> Approval of changes
> Trials
> Shelf life
> Labelling
> Nutrition claims

Clause 5.1.1

This is a new clause for Issue 6. It refers to situations where either you or one of your customers places restrictions on you regarding the introduction of certain hazards. It could be, for example, that a customer would not allow the presence of GM organisms on your site or certain allergens. In such a situation, you must provide clear guidelines regarding the scope of any NPD.

Clause 5.1.2

This is a requirement for all new products or any product or process changes to be approved by the HACCP team leader or other authorised team member. You will need to show the Auditor such an approval system. They might also ask for records of approval for any new products or processes introduced. Clearly, the approval dates must precede the introduction of products into the factory.

Clause 5.1.3

If you do carry out any NPD, you will always need to carry out factory production trials and thorough testing of new products or processes to validate that you can indeed make the product safely and to the required quality.

Example

A bakery specialising in pitta breads had introduced a new bread roll product. They also had a documented NDP procedure, which included the requirements for trials and testing. However, the Auditor found there was no evidence that the new product had actually gone through this procedure.
   A nonconformity was given.

   Additionally, you must also carry out such factory trials whenever you need to validate the process. This could be when you have made a change to equipment, for example.

Clause 5.1.4

This clause requires that your product shelf lives are established by taking account of the factory environment and storage conditions and handling of the product. The Auditor will be looking for evidence of these factors. This will be of special interest where your products are novel in some way. Even if your product is considered to be well established and have an 'industry standard' shelf life, you should be able to produce some evidence that you have ensured this. You are required to carry out shelf-life trials. The process must be documented, so too the results and the records retained (of course). Remember that you must consider microbiological, chemical and organoleptic criteria.

   Alternatively, if you produce something that has a very long shelf life such that trials are impractical, such as a canned product, you may use well-established literature as your source of shelf-life verification. Note that your references must be 'science based'.

Examples

At a manufacturer of cakes and doughnuts, the Auditor found that no final product end-of-life analyses or assessments had ever been performed.

At a manufacturer of beer, the Auditor was concerned to find that there were no results or other evidence documented to confirm the recent acceptability and justification for the shelf-life extension of all products from 12 months to 15 months.

Nonconformities were given in each case.

Note: See also Clause 5.5.1.3 for requirements on the continuing of monitoring of shelf life.

Clause 5.1.5

This is a new clause for Issue 6 that is concerned with legal labelling of products for the country of sale.

Note: This is not only about selling to a consumer but to other parts of the food supply chain, for example catering. You have to comply with this.

This is an important point because some seem to be under an impression that such requirements apply only to retail packs. They do not. Note that this includes ensuring that your product date coding system is correct.

Examples

At a bakery, it was found that there were incorrectly labelled shelf lives for white and brown sandwich loaves (5-day life as opposed to the correct 4-day life).

At another bakery, the company were found to have marked product with a 'use by' date when it should have been 'best before'.

A nonconformity was given in each case.

The Auditor will be looking at labels during the site tour.

The second part of the clause requires that all such labelling (ingredients, allergens and so on) match the product recipe. The Auditor might select a product (perhaps that used for the traceability test) and check recipe against label.

Example

At a bakery, there was no evidence of fresh batches of printed packaging being assessed for the accuracy of the printed information.
    A nonconformity was given.

Clause 5.1.6

The requirement concerns claims such as nutrition claims on labels. These are controlled by legislation in the EU (Regulation (EC) 1924/2006) and include the following:

- Low energy
- Energy reduced
- Energy-free
- Low fat
- Fat-free
- Low saturated fat
- Saturated fat-free
- Low sugars
- Sugars-free
- No added sugars
- Low sodium/salt
- Very low sodium/salt
- Source of fibre/protein/vitamins/minerals
- High fibre/protein/vitamins/minerals
- Contains X
- Increased X
- Reduced X
- Light/Lite
- Naturally/natural

There may be other claims for you to consider such as 'suitable for vegetarians' and the same systems must apply. The requirement here is that you validate your formulation and process to ensure that you meet any claims you are making.

The Auditor will determine if you are making any such claims and will want to see how you ensure that you meet them. It may be that recipes are studied at this point to see if your calculations, for example for reduced sugar, are correct. You should be prepared with evidence of validation. See also Clause 5.2.1.5 concerning allergen claims.

Clause 5.3: Provenance, Assured Status and Claims of Identity Preserved Materials

As mentioned previously in the previous issue of the Standard, Clause 5.2 included aspects of identity preserved products. It has now been separated, thus we have a new SOI for Issue 6. The SOI makes clear that traceability, identification and segregation are the keys to this objective.

What does this category include? The terms 'provenance', 'assured status' and 'identity preserved' all have definitions in the Glossary to the Standard and you should read them. Note that 'assured status' refers to a certification scheme.

The materials referred to could include organic, vegetarian, free range, GM free, kosher, Halal, Red Tractor and so on. They could also include products with a geographical status such as those with PDO or PGI. Note that these terms apply in the EU.

Should you not be handling such products, you may skip this clause.

Key points in Clause 5.3
Identity preserved or assured status claims
Traceability
Mass balance tests again
Preventing loss of control
Process flow document

Part Two

Clause 5.3.1

This clause is about any relevant claims you are making. You may be making product from scratch that will meet an assured status or you might be using such raw materials. Either way, where you are making such a claim on the product, your raw materials will be key. You must verify the status of raw materials and keep records accordingly. This may necessitate tight traceability systems on the part of your suppliers and yourselves and other documentation confirming the product status.

The Auditor will need to see this documentation.

Clause 5.3.2

This one is all about records and in particular traceability. Inevitably, when you are making claims about such products your traceability systems and record keeping are essential, but then they are anyway (see all of Clause 3.9). Of course, the Auditor will need to see that you are able to trace all such materials alongside, and as distinct from, any materials that you are not making claims about. In this case, elements like bin numbers and other containers take on special significance if they are dedicated items.

Note: In addition, you must also do mass balance tests at least every 6 months – more often if the particular scheme dictates.

Clause 5.3.3

Firstly, you must have a document that shows your process flows where you are making any such claims. Perhaps this could be tied in with other process flow diagrams (see Clause 2.5.1 or possibly Clause 4.3.2). Consider all parts of the site where such cross-contamination might occur despite diligent cleaning and so on, for example items of equipment that might be used for different types of product such as when you are making both meat-based and vegetarian products.

You must then ensure that you have controls in place to prevent loss of control here. The colour coding rightly indicates that this clause will be audited both in the site and as a document review and the Auditor will be looking closely at this aspect if you are making claims.

Example

A manufacturer of bacon-based and vegetarian dishes was found to have inadequate procedures to prevent contamination of vegetarian products with bacon.
A nonconformity was given.

Clause 5.4: Product Packaging

This SOI has two objectives, that packaging be suitable for the intended use and that it be stored appropriately. Both aspects are expanded on in the following clauses.

Key points in Clause 5.4
Information for packaging suppliers
Storage of packing
Product liners

Clause 5.4.1

You must have evidence available that your product-contact packaging is suitable for use. You may use certificates of conformity from your suppliers or other evidence for this. They must include all current references to legislation.

Example

At a produce packer, the certificates of conformity did not refer to current EU legislation (in this case EU 1935/2004).
A nonconformity was given.

Remember to include food contact materials such as bakery parchment.

Example

A bakery had no detailed specification for bakery 'parchment' (confirming its food-grade status) used to line unpacked breadbaskets.
   A nonconformity was given.

Of special importance, and new for Issue 6, is a requirement that you must inform your packaging suppliers of the nature of your food and any usage instructions such as microwaving. Clearly, features such as a high oil content in the food will have a bearing on the nature of material that can be used, thinking of possible migration of substances to the food, barrier properties, sealing properties and so on.

Clause 5.4.2

This is a simple requirement to store packaging away from raw materials and finished product, where appropriate. This is a matter of segregation and concerns any possibility of cross-contamination between packaging and food. Consider exposed food that might contaminate packaging and vice versa, whether by physical transfer or by taint.
   There is also a requirement to protect and identify part-used packaging material before returning to storage. Thus, reels of film, empty cans, jars and so on must be covered. This is a procedural matter that staff should be trained in.

Examples

At a wine bottling plant, the Auditor found that not all part-used pallets of bottles were adequately covered or protected.

A bakery was found to have an amount of uncovered, part-used packaging film in storage.

At a poultry plant, part-used contact liner packaging was stored in the warehouse area in designated containers. However, these containers were not fully protected and have part-open lids.

Another bakery had part-used batch of fruit salad primary packaging tubs that where not stored in a protected manner, notably face up with the upper tub exposed.

Nonconformities were given.

Remember also to isolate obsolete packaging so that it cannot be inadvertently used. The Auditor will look at storage during the site tour.

Clause 5.4.3

This requirement concerns product liners. This is usually plastic material such as polyethylene and the point is that it shall be coloured. This is to reduce the chance of product contamination; clear plastic is less easy to spot in product.

The requirement is for all stages including raw materials, so you should specify to your suppliers that they too use coloured material. The requirement does not specify which colour but it should be a different colour to the food itself. For that reason, blue is the best in most circumstances as there are few blue foods. However, understandably a manufacturer may use different colours to distinguish, say, low- and high-risk products.

Example

At a poultry manufacturer, the Auditor found clear plastic film used as liners as well as blue film.
    A nonconformity was given.

The material must also be of a sufficient gauge. I well remember a number of complaints of blue plastic in product caused by a manufacturer who was using fine blue polythene to line large bins of raw meat after butchery. The material was evidently too thin and the problem was only resolved when they switched to a thicker grade.

This clause may also apply to paper materials, for example baking sheets.

Example

At a bakery, holed silicon sheets were being used to line cookie baking racks. Clearly, this was a contamination risk.
    A nonconformity was given.

Clause 5.5: Product Inspection and Laboratory Testing

This SOI is an objective to carry out inspection and analyses appropriate to your product.

Key points in Clause 5.5
Testing schedules
Reviewing results and trends

Monitoring shelf life
Sensory analysis
Laboratory standards
Positive release

The Auditor will want to see results of all analyses critical to product safety, legality and quality and will look to see that such work is carried out to a reasonable frequency. This may be carried out in house or by external laboratories.

Examples

At a sausage manufacturer, it was found that there were no results available for the any meat content analysis for the previous 18 months.

A supplier of chocolate products was found to have carried out no chemical laboratory analyses of final products during the previous 12 months, notably for milk solids, cocoa solids and butterfat.

Nonconformities were given in each case.

Note that the SOI also indicates that any such procedures must be carried out themselves in a way to prevent any risk to product safety. This would include on-site laboratory facilities and the methods used in them. The Auditor will want assurance that you take all relevant precautions.

Part Two

Clause 5.5.1: Product inspection and testing

Clause 5.5.1.1

This requirement concerns testing and inspection schedules. Firstly, you must monitor compliance with specification by testing and inspection. Secondly, the procedures must be documented, that is test methods and frequencies.

Therefore, the Auditor will be looking for a risk-based system that shall be documented. The procedures should include what to look for and how, specifying tolerances. A typical example would be moisture checks on dried products. In that case, the Auditor would want to see instruction on how to use the instrument, how often to do the test, records of checks, limits and tolerances and instruction on what to do if out of specification.

The clause states that your testing programme may include organoleptic testing. This should be the norm in fact unless your products are not consumer products, for example food ingredients such as additives. The Auditor is very likely to want to see evidence of this. Consider the frequency of sensory testing. This is something you might risk assess based on the natural variability of the product.

> **Example**
>
> A bakery was unable to provide evidence of organoleptic assessments of fresh final products.
>   A nonconformity was given.

Note: The requirement for testing includes the production environment.

Clause 5.5.1.2

This clause concerns the use of test results and there are two aspects. Firstly, you must record and review your results and identify any trends. Results are usually recorded but the review of results is sometimes overlooked. The Auditor will want to see evidence of review, so it will not be enough to present the Auditor with a sheaf of laboratory results with no sign that these have been read and understood and analysed for any trends.

Secondly, you must show some action in the event of results that are out of specification, especially where a trend has been shown.

> **Example**
>
> A manufacturer of meat products had regular samples sent to an external, accredited laboratory, including product samples and environmental swabs. No corrective actions had been documented when a positive *Listeria* result occurred.
>   This is a surprisingly common oversight and as usual in these instances a nonconformity was given.

Clause 5.5.1.3

The requirement concerns the ongoing assessment of product shelf life (see also Clause 5.1.4 concerning the establishment of shelf life). You must continue to validate shelf life, and your system for doing so shall be based on risk. The clause details the parameters that you must consider, microbiological, chemical and sensory, and that you record these tests.

This clause states that you must validate the shelf life stated on the product. You may need to provide facilities for the retention of samples. Some products have a very long shelf life such as canned goods. This might be a challenging requirement in that case and you may be able to justify that such long life packs are very low risk in terms of an issue with shelf life. For foods with a mid- to short life and especially those that are perishable, the Auditor will undoubtedly expect to see records of ongoing checks.

> ## Examples
>
> A manufacturer of cakes and muffins was found to have carried out no final product end-of-life assessments and shelf-life confirmation undertaken in the previous 3 years.
>
> At a fruit processor, there were no final product samples retained for 4 kg frozen concentrate and therefore, no ongoing end-of-life assessments performed.
>
> A manufacturer of grated cheese and processed cheese, packed under modified atmosphere, was found to have no ongoing, final product end-of-life analyses/assessments performed at all.
>
> Nonconformities were given.

Clause 5.5.2: Laboratory testing

Clause 5.5.2.1

This is a straightforward requirement concerning pathogen testing. You must either send samples off site by sub-contracting pathogen testing to an external laboratory or if you carry it out on site, you must separate the function completely from production areas by setting your laboratory facility remotely. It is likely that the only way to achieve such remoteness is by having a separate building, although it could be in the same building provided there is effective segregation that prevents any risk to the products. Therefore, you have also to ensure that your procedures prevent risk to product.

   Not many companies carry out their own pathogen testing these days for obvious reasons.

Clause 5.5.2.2

It might seem from reading the requirements under Clause 5.5 that there is a wish to discourage on-site laboratories. Not so, of course. This requirement, however, concerns the possible risks involved in having on-site laboratories for routine testing and in particular their location and design. The clause lists six bullet points that shall be taken into consideration. Firstly, you must consider the design of the laboratory including drainage and ventilation systems. The key factor is that nothing going on in your laboratory must be allowed to cause contamination to the products.

   The security of the laboratory and access of personnel, including the movement of laboratory staff, are important to consider, so too are any necessary arrangement for changing clothing. It will be expected that staff do not move from a microbiological laboratory without changing overalls. Sample collection must also be carried out with due regard to product safety and of course to the accuracy of results.

   Laboratory waste is a very important consideration and there must be proper arrangements for disposal. Note also that you might need to include laboratory glassware in your glass register.

Part Two

The six points required in this clause must form part of your documented controls.

The Auditor will have checked the site plan for the location of any laboratories including the drain routes and flow of staff to and fro and will look at laboratory practices during the site tour. Your documented controls must also be available for examination.

Clause 5.5.2.3

This has caused a little confusion in the past. Put simply, this clause requires that where analysis is carried out to ensure product safety or legality (not quality), the laboratory carrying out the work shall either be accredited to ISO 17025 or operating to that standard anyway. This is the case whether you are doing the work or sub-contracting it to an external laboratory.

Firstly, consider all the tests that are being carried out either in-house or externally. If any of them relate to food safety or legality then they must conform to this. Many factories have taken the view that it is easier to meet this by sub-contracting the work to an accredited laboratory, but a good number still do their own work.

If you are using an outside laboratory, there is only one way to ensure complete compliance in the eyes of the Auditor and that is by making sure that they are accredited. There is no realistic way that the Auditor could ascertain that they are working to the principles of ISO 17015 without being so accredited unless you have undertaken some serious auditing work yourself. I have never come across that situation.

Considering accreditation, it should preferably be by a national body such as UKAS in the United Kingdom, COFRAC in France, Accredia in Italy and so on. As it happens, there are some other private schemes in the United Kingdom that are also acceptable and an FTCG Position Statement on the subject stated that 'laboratory accreditation to 17025' or any other recognised laboratory scheme, for example 'LabCred' will satisfy this requirement'.

If you are carrying out such testing in-house the Auditor will want to see evidence that you meet the requirements of ISO 17025. The Auditor does not have time to carry out a full audit of your laboratory so you must be able to provide some readily accessible evidence of conformity. A certificate of accreditation would be excellent, otherwise this might include:

- staff training (see also Clause 5.5.2.4);
- written methods;
- standardisation or calibration of laboratory equipment and chemicals;
- ring testing with an accredited laboratory; and
- actually having a copy of ISO 17025 on site!

The aforementioned Position Statement also says that 'If an external laboratory of repute is verifying results then this will also be acceptable'.

Example

A bottler of mineral water was using an external laboratory for critical microbiological testing. The Auditor could find no evidence that they were either accredited or otherwise operating to ISO 17025.
  A nonconformity was given.

Note also that accreditation should include all the tests being carried out. The Auditor will study any test results from accredited laboratories and be on the look out for any disclaimers for tests that are not within their scope of accreditation.

Example

A manufacturer of frozen sweetcorn was found to have no documented confirmation that their external laboratory's scope of accreditation included GM analyses.
  A nonconformity was given.

The clause requires that where accredited methods are not used, you must justify and document this.

Clause 5.5.2.4

This again has been the cause of a little confusion, referring as it does to results other than those specified in Clause 5.5.2.3. What this requirement refers to is tests carried out on the quality of product, not food safety or legality. For such tests, there is no requirement for accreditation or working to ISO 17025, but you must ensure the reliability of results.

The clause lists five points that you must include in your procedures such as ring testing to verify the accuracy of results. Consider all the tests that you do.

Example

A soft drinks supplier was testing for total solids and acidity; however, the Auditor found that there was no procedure in place to ensure reliability of the testing techniques being used for °Brix and acidity.
  A nonconformity was given.

Part Two

One of the specific requirements is for analytical personnel to be suitably qualified and/or trained and competent. For the accredited external laboratory, the Auditor will take this as a 'given'. For the internal laboratory, the Auditor will look for evidence of qualifications, for example certificates and training or staff appraisal. Often, the Auditor will take note of an individual or two in the laboratory and look at their staff records when also covering general training under Clause 7.1.

You must be using recognised test methods and your testing procedures must be documented. The Auditor will ask to see examples of these. Finally, the testing equipment used must be calibrated and maintained where appropriate, so have some relevant records available (some equipment will require standardisation rather than calibration).

Clause 5.6: Product Release

This SOI is an objective that procedures are not bypassed in any way and that unauthorised product cannot be released. There are many factors for the Auditor to consider here including your procedures, factory layout and staff training, and the Auditor may only conclude whether or not this requirement is met by having seen all processes in action. Overall, the Auditor must be satisfied that you display control over finished product. Conversely, nonconforming product must be controlled (see Clause 3.8).

> **Key points in Clause 5.6**
> Controlling release of finished product
> Positive release

Clause 5.6.1

In certain industries, positive product release is the norm. For example, some manufacturers might carry out microbiological testing before releasing batches. In canning, batch samples are usually incubated at elevated temperatures. They are then checked, for example for pH, before release. Note that release must be authorised in this case. The Auditor will want to see evidence and records in such cases.

Example

At a mushroom canner, product was positively released on incubation results. However, the way of recording incubation result was poor on detail. Incubation temperature is not monitored.
A nonconformity was given.

Summary

Section 5 deals with some general subjects such as product testing and packaging and some very specific subjects such as allergens. All subjects concern control though, which in turn means having good documented procedures that are well thought through and good staff awareness through work instructions and training.

Quiz No. 10

(1) How often must you carry out mass balance testing on provenance, assured status or identity preserved products?
(2) You do not need to worry about allergen procedures because you have comprehensive warning statements on labels: true or false?
(3) When do you need to ensure that the laboratory is operating to EN17025?
(4) What must you do when specifying food contact packaging?
(5) How many clauses in Section 5 specify a training requirement?

Part Two

16 Clause 6: Process Control

Clause 6.1: Control of Operations (Fundamental)

This is a *Fundamental* SOI and your objective is to ensure product safety, legality and quality by the use of documented procedures or work instructions. Naturally, they must reflect your HACCP plan.

> **Key points in Clause 6.1**
> Correct work instructions
> Controlling your processes
> Variable process conditions
> What to do when equipment fails
> Pre-production checks
> Handling product changes

Clause 6.1.1

This requirement is about making sure that work instructions and product specifications are available to those that use them, and, of course, that they must be correct.

Example

At a speciality bread bakery, the Auditor found that processing parameters for prover relative humidity had been changed in the process area from parameters set by management, without management knowledge, consent or approval.
  A nonconformity was given.

The clause lists no less than eight bullet points that the specifications or work instructions must include if they are relevant to the system. They include obvious facets of production such as cooking, cooling, labelling, etc. Anything not specifically included will come under your CCPs anyway (final bullet point) such as pack seals. Remember that product recipes and dosage rates must be adhered to.

> ## Example
>
> At a vegetable processor, there was some sloppiness in recipe control as follows. There was no confirmation that the Dry Wite (potato preservative) was prepared accurately to the correct defined concentration (250 g Dry Wite/250 L water) as the operator did not weigh the Dry Wite but used a 'cap-full'.
>
> At the same site, there were no checks to confirm that mixed vegetable or fruit products are being prepared to the correct proportions (as defined in recipes).
>
> Two nonconformities were given.

Question: If my product is mostly one ingredient that is an allergen such as milk in cheese, do I have to emphasise 'milk' in the recipe?

Answer: If you only make cheese and thus the allergen is the major ingredient in all products there is not an absolute requirement to highlight it in the recipe.

Note: The Auditor may check the content of the documents during the document review but the clause is also about availability and is colour-coded peach to indicate that it will be checked during the site tour.

Clause 6.1.2

This brief clause is a very important one, which covers all aspects of your process and how it must be controlled. It refers to temperature, time and pressure control, but, of course, this could also include humidity, pH and mixer speeds; there are many possible measures. For example, for a cooking and cooling process, you must verify that correct temperatures and times are achieved. In a cannery, for example, it would be expected that heat penetration is checked such that correct F_0 or F_{100} values (as appropriate) are confirmed. Commonly, this is done using data logger systems. In pasteurising processes such as for milk, the temperature and holding times should be checked and verified. In such processes, safeguards such as divert valves, which operate when temperatures drop too low, would be expected, and the operation of the divert valves themselves must also be verified. In other cooking processes such as roasting or baking, the checking of core temperatures would be a necessity.

> ## Example
>
> A bakery had no defined and measurable parameters (e.g. post-oven and pre-packing product temperatures) for monitoring the baking or cooling process steps of bakery products (they were currently relying on visual and touch).
>
> A nonconformity was given.

There are many other examples of measurable parameters. Bear in mind that this does not apply only to cooking and cooling but any process affecting the safety, legality and quality of product. Thus, drying, concentrating, chilling, freezing, mixing, blending, cutting and so on should all be verified. For instance, in the drying of pasta, humidity control is vital to ensure that drying is uniform. Verification might also include product sample testing as well, thus microbiological testing, incubation testing, pH, moisture content and other analyses may also play a part in this.

Consider your entire operation from raw material intake through to finished product dispatch. Thus, storage conditions are also relevant.

The clause requires that such aspects are established, monitored and controlled. The Auditor will be looking at this aspect during the site tour and during document reviews and will want to see evidence of this. Records must be accurate and show a process under control.

Examples

At a bakery making sandwiches, the Auditor found that the chiller in the low-risk section of the sandwich production unit was operating above 10°C.

A pies and quiche manufacturer was found to have one refrigerator operating at 13°C although recent reports showed it to be at 5°C.

At a high-risk products plant, it was found that there was no temperature monitoring of the sauce once it had started to be packed. For example, on the day of the audit, it was 63°C at 15:30 but was still being packed at 17:00 with no further monitoring.

A low-risk bakery had no documented validation that the oven air temperature settings will result in the correct baked product core temperature (this is required as post-baking core temperatures are not monitored).

At a tomato cannery, the thermograph on the 3-kg line pasteuriser had not been changed from previous day.

At a manufacturer of Indian meals, several problems were noted. Cooked, bulk-packed chilli beef was left out of refrigeration during a staff break; the sauce base cooking/cooling on-line records were unclear as to whether cooling times were adequate to ensure product safety.

In a tomato cannery, the Auditor found that there was no continuous temperature record on the 500-g can pasteuriser. Also, the thermograph on the 3-kg pasteuriser was not recording due to a defective pen.

A manufacturer of frozen ready meals was found to have no continuous monitoring of the freezer.

A wine manufacturer was found to have no definition of the temperature range for thermostatic wine storage tanks.

A fresh meat products site had a central system for monitoring temperatures in all production and storage areas, which produced a printout. The Auditor noted that both date and time were incorrect on the printed records.

Part Two

> At a high-risk site, the Auditor found no reaction or corrective action documented when chilling time exceeded the specified maximum 4 hours.
>
> Nonconformities were given in each case.

Clause 6.1.3

It is for you to arrive at the most appropriate settings for your process and product characteristics. This requirement concerns in-line monitoring devices such as temperature recorders that give continuous records and where staff are not observing the temperature continuously. To meet this requirement, it is vital that where you have in-line monitoring devices there is also a failure alert system. This failure alert must also be tested.

Thus, in a typical low-acid cannery, the Auditor will expect to see your capability to monitor the process, the failure alert or alarm systems and records (e.g. conveyor speed, time in retort, temperatures, air and steam pressures, etc.) for each batch of product. In a typical liquid pasteurising process such as milk, ideally, chart records showing continuous time, temperature and divert valve checks would be the norm. Audible or visible alarms would also be expected in the event of process failure. In a continuous baking operation such as for bread, oven temperatures, baking times and the measurement of core temperatures at oven and cooler exits would be expected. The computerisation of such processes has advantages, for instance the control and monitoring of pasta drying is greatly improved with display screens showing the various zones of different temperatures and drying humidities.

The Auditor will want to see such systems in place both during the tour and from your records. Paper chart records should be clearly dated with the product indicated. Any computerised records should be available for retrieval and it should be possible to trace individual batches. These records should show your tests of alert systems.

Testing the failure alert is straightforward in systems like liquid pasteurising where the divert valve system can be checked by artificially lowering the process temperature. In other systems, you may also have to artificially change the process conditions in order to fire off the alert.

Clause 6.1.4

This clause is a new clause for Issue 6. It relates to situations where there is variation within processing equipment such as hot spots or cold spots in ovens or freezers. This is possible in ovens, for example, especially where they are large and continuous. In such cases, you must validate the consistency of the process. You might not need to do this every day but at a frequency based on risk. If this applies to you, have something ready to show the Auditor and be ready for a challenge if you have done nothing at all about this. The Auditor should be aware of inherent variation in your technology.

Clause 6.1.5

This requirement concerns what you must do if the process fails while product is still within the process but cannot be diverted automatically. Thus, if a continuous cooker breaks down you must have contingencies in place to deal with the product that is sitting inside it, while you take action on the equipment. You must have procedures to deal with the release or otherwise of such product and may have to take into account the point at which failure occurred, the amount of downtime, the microbiological status of the product and so on. There must be a system in place to determine whether, after the failure is corrected, the product might be safely released to continue the process or whether it must be discarded, which in turn might invoke a nonconforming product procedure (see Clause 3.8).

The Auditor will be looking for a controlled approach to this and you should be ready to answer some 'what if' questions.

Example

A baker of baguettes was found to have confusing procedures in the event of equipment failure. Procedures for equipment failure and so on did not satisfy the requirements of the clause, they were a combination of a number of other procedures and were not clearly stated.

A nonconformity was given.

Part Two

Clause 6.1.6

This clause is about checking production lines at start up and at product changes. You should be checking for hygiene and to ensure that packaging has been changed accordingly. Of course, these checks must be documented. Note that this clause is colour-coded both green and peach and it accurately indicates that the Auditor will check conditions during the site tour when changeovers happen, but also might ask to see records during the document review or vertical audit.

Note: There is a close link between this clause and Clause 4.11.4.

Clause 6.1.7

This clause is about having procedures to ensure that the correct packaging and labels are used at all times including when you have product changeover. This can easily happen on high-speed packing lines but problems can also occur in other circumstances.

> **Example**
>
> A packer of coleslaw used plastic tubs with hinged lids and a large sticky label that was intended to fold over the open edge opposite the hinge and give tamper proofing. On the machine, it was impossible for labels to be fed on the wrong way round but as tubs passed from the labeller if any had missed a label they were hand fed back through. Thus, any tubs put on the wrong way round were labelled over the hinged edge. This indeed was happening. *Result*: Tubs that were not tamper-proof and in fact could flip open at any time.

Incidents can occur when product is being changed over but the labels are not changed effectively.

> **Example**
>
> A canner of chopped tomatoes caused a complaint with a UK local authority where they changed over from chopped tomatoes with garlic to chopped tomatoes with chilli. A small amount of cans containing chilli were labelled with the garlic product label and found their way into the UK market.

Note that you must have checks right through from the start of any run, during a run and following any changes of packaging or packaging materials. You must also check coding and any other information printed by you on the line.

Clause 6.2: Quantity – Weight, Volume and Number Control

Weight and volume control are subject to clear legislation and in some cases additional codes of practice. This SOI requires that you meet legal requirements, notably where the product is sold. In other words, you must be aware of any particular legal requirements in any countries that you are exporting to.

Key points in Clause 6.2
Legal requirements in all relevant countries
Checking regimes
What to do about bulk items

Regarding retail and catering size packs, UK legislation on weights and measures was revised in 2006, although much of the essential principles, such as meeting the three packer's rules, remained the same as before (The Weights and Measures Packaged Goods Regulations, 2006). This law implements various EU directives, which in turn apply throughout the EU.

I would just like to add a note here concerning the 'e' mark that is applicable to product packed to this EU system. The system that applies throughout the EU allows packers to pack product to meet the nominated quantity on average. This means that individual packs may be below the declared quantity provided the whole batch meets the rules, one of which is that the whole batch meets the declared quantity on average. I have sometimes seen the 'e' mark misused. It need only appear on products that are being moved between EU member states. It is not required on product that is sold only in the country of manufacture (although it does not hurt) and it definitely does not apply to products made outside the EU.

In other parts of the world such as Canada, they also have a similar system whereby packs must achieve an average net quantity with certain tolerances. In the United States, though the situation is different in that it depends on the state legislation, commonly a minimum quantity is required (see http://www.inspection.gc.ca/english/fssa/labeti/retdet/bulletins/nete.shtml).

Most manufacturers will use non-destructive testing and the use of automated check-weighing systems. You should consider the frequency at which they should be tested. The requirement also mentions customer-specified requirements. This might include quantity numbering on retail packs; also, bulk items are likely to be relevant to this.

The Auditor will also be looking for a controlled system that covers all your products and any countries into which you are selling. You shall need to show awareness of the legal requirements of any such country that you export to.

Examples

A wine bottler was found to have no checks or records of weight/volume control of filled bottles.

A confectionery manufacturer was found to have no records maintained of minimum weight control.

A nonconformity was given in each case.

The precise detail of your systems is covered in the subsequent clauses.

Note: Because quantity comes under legal requirements, a failure here could result in a critical nonconformity.

Clause 6.2.1

This requirement covers the detail of the checking of quantity and that it must meet legal requirements for method and frequency. The Auditor will want to see records of checks. They may be manual checks taken on-line or they may be printouts of automated systems. During the tour, the Auditor may also ask to see a test being

Part Two

carried out. In the case of automated systems, the Auditor will also want to see that they have been challenge-tested, for example to see if a short weight pack is rejected.

In the case of product being sold to a minimum quantity, the Auditor will want to see systems that demonstrate that every pack meets the quantity requirement. In the case of packs sold to an average quantity, the records must show that sampling rates and frequencies meet requirements and show any action taken when the relevant rules are not met. The following examples all refer to product sold in the EU.

Examples

A manufacturer of liquid and dry food flavours and ingredients had no system to confirm and monitor that pack weights were being achieved correctly (all packs were to minimum weight or volume).

At a speciality bakery, oatcakes were being packed to a minimum weight of 200 g; however, the on-line documentation was found to have a tolerance of 10 g.

A confectionery manufacturer was found to have no record maintained of the testing of the in-line check-weigher automatic T2 rejection mechanism.

The recorded final product weight checks for speciality breads packed during the early morning on a specific date were consistently approximately 30 g below the minimum acceptable weight; however, no corrective action had been taken to quarantine these products or prevent them being despatched for sale to the customers.

A company that was cutting, slicing and grating cheese was found not to be calculating mean weights for each batch of grated cheese.

At a pies and pasties manufacturer, the Auditor noted that while the procedures stated that weight checks were carried out at a rate of five packs per $\frac{1}{2}$ hour, they were being carried out hourly.

At a baker of cakes and muffins, the Auditor found that underweight final products were not rejected into the locked cage, due to the reject cage being over-filled with rejected products. Also, final product multi-bags containing less than the stated number of cakes were not rejected (this was because the total pack weight met the minimum weight requirement).

A baker of pitta bread was found to have operator misunderstanding of the correct procedure for procedure for recording final product weights and subsequent corrective actions when underweight product identified. Also, weight control parameters for final products were not adequately available to operators.

At a manufacturer of dry breadcrumbs in large bags, it was found that weight control documents did not reference upper and lower tolerance limits. The weight control documents did not indicate any corrective actions if a low weight was recorded nor did they reference the accuracy limit between the machine reading and the manual scales.

Nonconformities were given in each case.

As you would expect, you must keep records of all checks.

Clause 6.2.2

This requirement is concerned with quantities that are not covered by legislation such as bulk materials. In such cases, you must have customer specifications for quantities and systems in place to demonstrate that you meet the specifications. The Auditor will want to see quantity control and records, for example for bulk tankers, containers, tote bags and so on.

Clause 6.3: Calibration and Control of Measuring and Monitoring Devices

Your objective in this SOI is to demonstrate that all measuring devices and monitoring equipment are accurate and reliable. A large part of this will concern calibration.

Key points in Clause 6.3
Identifying and listing the relevant equipment
'Master' instruments
How often should you calibrate?

This SOI concerns product safety and legality in that all such measurements, be they CCPs or otherwise, must be done using identified equipment. Furthermore, where appropriate, such equipment must be calibrated to a recognised national standard.

The Auditor will be aware of your CCPs and any other measurable aspects that are critical to food safety and legality and will want to see evidence that all such equipment is calibrated or standardised.

Part Two

Examples

A manufacturer of chickens and chicken portions was packing into modified atmosphere packs. The Auditor found there was no in-house calibration performed, or records maintained, for the oxygen analyser (between external calibrations).

A manufacturer of pastry confectionery was found to have no calibrations of the integral temperature monitoring system for chilled or frozen raw material storage facilities.

At a manufacturer of sliced and grated cheese, it was found that there was no record of calibration of the probes used within the factory.

At a pasta factory, it was found that there was no evidence of thermometer calibration for the two incubators used for microbiological analysis.

Finally, in an unusual example, test weights used to verify the bakery scales were not the company-issued test weights, but test weights owned by an operator (and stored at his home).

Nonconformities were given in each case.

Clause 6.3.1

This is a requirement concerning only equipment used to monitor CCPs. You must identify and control such equipment. All such equipment must at least bear an identifying mark, either a company code or the manufacturer's serial number. You must also keep a list of such equipment and the list must include the equipment location.

The Auditor will ask to see your list.

Commonly, companies attach stickers that state equipment numbers and give the calibration status and indicate when the next calibration is due. The Auditor will look at all such equipment during the site tour and will ask to see certificates or other records that match the identifying marks or serial numbers.

Examples

At a confectionery bakery, it was found that there was no calibration status identification on the thermometer and hygrometer on one of the freezers.

A canner of pulses was found to have no evidence of the calibration status or identification of measuring instruments (e.g. thermometers on retort control panel).

At a manufacturer of processed cheese, the Auditor found that no calibration was currently undertaken of the air temperature probe (used to monitor certain CCPs).

Calibrated equipment is listed but calibrated equipment is not identified by tags and due dates are not documented.

In an instrument list, only laboratory instruments were listed and no process instrument (e.g. thermograph and temperature regulator).

A bread bakery was found to have temperature probes that were not marked or identified in any way.

At a manufacturer and bottler of cream liqueurs, it was found that some of the mercury thermometers were not clearly identified.

Nonconformities were given in each case.

You must also prevent their adjustment by unauthorised staff. The Auditor will want to see evidence of who is responsible for such equipment and accountability where any adjustments are made. Typically, certain appointed staff will be responsible for the adjustment of, say a pH meter. Records should indicate who has made any adjustments and the Auditor will want to see such records.

Similarly, this requirement also demands that such equipment is protected from damage, deterioration or misuse. For this and for the previous clause, the Auditor will be looking to see how the equipment is handled and stored during the site tour and generally expect to see a good level of care of sensitive instruments.

Examples

I was once at a bulk packing site and found that they were using a 5-kg test weight as a doorstop.

In another example, the 5-kg standard weight used to verify the weighing scales was uncalibrated and severely corroded.

A nonconformity was given in each case.

Clause 6.3.2

This requirement is about the frequency of checking of equipment. The frequency must be pre-determined. Where CCPs are measured with instruments, they must be calibrated at an acceptable frequency.

Example

A fruit juice processor had allowed their frequency of calibration of their pasteuriser to lapse: no record of calibrating the pasteuriser temperature probe (PT100) since early in the previous year.
    A nonconformity was given.

The Auditor will want to see the schedule of calibration, standardisation and so on as evidence of this. Also, the accuracy of devices shall be within agreed parameters.

Examples

A blender and packer of oils was found to have no documented tolerances for in-house calibration of final product scales.

A cheese and dairy products maker was found to have no defined calibration parameters for temperature probes or scales, when calibrated in house.

At an apple cannery, it was found that not all instruments (some new) were calibrated at the planned frequency.

Part Two

> At a produce packer, it was found that the scheduled chill room probe units calibration had lapsed.
>
> At a poultry processor, it was found that the calibration frequencies for scales and weights were not documented.
>
> Nonconformities were given in each case.

The methods used must be traceable to a national standard. If you are using a contractor or if you are carrying out checks yourselves, the Auditor may ask for evidence of the methods used.

It is vital to keep records of calibration and these must be available for the Auditor.

Examples

A wine bottler was found to have no calibration records for the incubator thermometer, no recorded checks on alcoholometer (certificated reference solution) or any records for identification of calibration status on tank thermometers.

At a manufacturer of potato products, it was found that there was no third-party calibration record (i.e. traceable to a national standard) available for one of the oxygen analysers during the previous 12 months. Also, there were no records maintained of the daily in-house calibration of the oxygen analysers.

A poultry processor was found to have no records maintained of the external service/calibration of the critical integral temperature probes, that is within chillers, freezers and smokehouses.

A milk dairy was found to have no calibration records available on site for the torque meter.

A potato products manufacturer was found to have no external calibration certificate available for the 25-kg packing line final product scale.

A produce packer was found to have no procedure in place to record calibration of weigh cells and 'on-line' check-weigher.

A snacks manufacturer was found to have no record available of the previous year's in-house calibration of hand-held temperature probes.

Nonconformities were given in each case.

Finally, note that your equipment must be readable and be of suitable accuracy. This means that the resolution shall relate to what you are measuring. Do not go using a 5-kg scale for checking a 20-g item.

Examples

A jam factory had no 5-kg test weight to verify the calibration of the final product scale (notably, for the 4-kg frozen concentrate products).

In another case, the laboratory scale was not appropriately calibrated (range 0–5000 g, calibrated 10–1000 g, field of application 200–3500 g).

A nonconformity was raised for both.

Clause 6.3.3

This clause refers only to reference equipment. This could be a master thermometer from which you calibrate other probes, or it could be a master set of standard weights, for example. Any item like this must be calibrated to a national or international standard. This is a very important clause and the concept of calibration being traceable to a national standard must be understood, especially if you are carrying out calibration internally. For example, to calibrate a balance using standard weights is fine but the weights themselves must be calibrated to a national standard. In such circumstances, the Auditor will want to see evidence that the company carrying out the calibration can demonstrate that their measurements are traceable to a national standard.

If the external company used is accredited, for example in the United Kingdom by UKAS, then this will confirm traceability. However, many contractors are not accredited. Therefore, if the contractor is not accredited, you may have to provide evidence of traceability.

Remember also that you must calibrate such instruments for their normal working range.

Examples

A manufacturer of cooked meats told the Auditor that they had a new master temperature probe. However, it was not available for examination and there was no evidence that it was adequately controlled.

A jam manufacturer was found to have no calibration certificate available for the test weights used to calibrate their scales.

At a bakery, the reference temperature probe was not calibrated across its range of use notably −18°C and 0°C.

A nonconformity was given in each case.

Part Two

Clause 6.3.4

This requirement is for procedures and for records of action taken. You must record when calibrated or standardised instruments are outside their operating limits and what action you have taken. The Auditor will expect to see defined procedures for this and ask to see records of any instances where such actions have been necessary.

> ## Examples
>
> A wine manufacturer was found to have procedures for calibration, however, corrective actions when instruments were found not to be operating within specified limits were not defined in the procedure.
>
> At a baker, the Auditor found that the probe 3 calibration records indicate that the unit was outside parameters on the last two monthly checks. The out of calibration reporting procedure was not implemented or actioned.
>
> A pasta factory was found to have no procedure to record actions taken when instruments and equipment are found to be out of calibration.
>
> A nonconformity was given in each case.

You must always be aware of the possibility of equipment becoming inaccurate. If it does and in consequence product no longer conforms in terms of safety or legality then you must take action to prevent such product being sold.

Summary

The requirements packed into the shortest section of the standard are, of course, as vital as all the rest. It concerns parts of your business that you should know better than anyone: how to control your processes to ensure food safety, legality and quality.

> ## Quiz No. 11
>
> (1) What two checks must you make on your production line before starting production and at product changes?
> (2) What equipment must be calibrated to a national or international standard?
> (3) If you use in-line monitoring what must the system include?

17   Clause 7: Personnel

This section is about all matters concerning personnel. Staff are always a slight worry to most Technical Managers because of their very human tendency to being unpredictable. Good training, supervision, work instructions and facilities should go a long way to reducing the worry.

> ## Example
>
> Many years ago, in the days when UK retailers carried out their own technical inspections of their suppliers, I was helping a new pasta factory (pastificio) get ready for their first visit. The visit was by someone who I respected very much, and of course, I was keen to impress.
>
> Being new, the factory itself looked wonderful. It was spacious and well laid out with plenty of room for expansion. The personnel were well presented with new coats and hats and I knew that the site was state of the art at the time and literally the newest factory in the country. The equipment shone.
>
> This particular location did have a reputation for staff smoking in those days but we had done much (so I thought) to eliminate the problem. Imagine my embarrassment when, during the site tour, this wily inspector poked his pen into a round hole in one of the main pillars bearing the roof and pulled out dozens of cigarette ends. He then repeated the process at each of the pillars in turn, leaving little piles of butts by each.

Clause 7.1: Training – Raw Material Handling, Preparation, Processing, Packing and Storage Areas (Fundamental)

> It's all to do with the training. You can do a lot if you're properly trained.   (Elizabeth II)

Training is fundamentally important to your systems, so, guess what: this SOI is a Fundamental Requirement. You must be able to demonstrate to the Auditor that systems of training are well established and the Auditor must be able to see that staff are adequately trained, supervised and competent from what they see of their activities and your training records. It is also important to have records of staff qualifications, including any that were obtained externally.

Key points in Clause 7.1
Induction
CCP training
Assessing competency as well as training
Procedures and records
Training review

Clearly, your training systems must be firmly established and part of your culture. You must also consider all languages spoken on site. The Auditor will note if a site is multilingual and ask what provision you have made in your training for all languages.

Remember also that we noted in Chapter 4 that there are a number of clauses that require specific training in certain areas. Those areas are internal audit, complaint handling, site security, chemical control, cleaning, pest control, allergens and laboratory testing. The Auditor might well return to these subjects when covering training as a whole.

Example

At a vegetable canner, training records were kept. However, those for site security, housekeeping and pest control were not in place.
A nonconformity was given.

Clause 7.1.1

This requirement concerns training prior to commencing work. Most companies have what is referred to as an induction for new staff. In order for the Auditor to assess this, you must keep records of induction and it must include all aspects necessary for the integrity of your product. For example, it would be expected that all new staff in the relevant areas receive some instruction in your Quality Policy, food hygiene rules, management structure and reporting lines. In addition, as well as general induction, there must be specific training for the actual job.

Note that this requirement also refers to temporary staff and contractors. In other words, you must treat them the same as permanent staff in this regard. Training needs will be determined only on the nature of their activity, not their status.

Example

A fresh meat processing plant used hygiene contractors. In the circumstances, the Auditor considered that the contractors had not received any training on induction and in this way were not sufficiently trained in the same way as permanent staff.
   A nonconformity was given.

The Auditor will examine training records to establish these points and may also speak to staff during the site tour to gain insight into the adequacy of their training. A typical question would be 'Are you aware of the company Food Safety and Quality Policy?' for example.
   The clause also requires that all personnel are adequately supervised while working.

Clause 7.1.2

This is a more specific requirement concerning personnel who are involved with CCPs. Such staff must receive any necessary training to carry out the CCP-related activity. Also, you must document any monitoring of such staff.
   Clearly it is essential that all staff involved with CCPs, such as carrying out a monitoring activity, are properly trained for that activity. The Auditor will ask to see records that specifically relate to this and may ask relevant staff during the tour how they were trained.

Examples

A speciality bread bakery was found to have no evidence of CCP training available for relevant staff.

A cereals milling company had documented procedures for the training of CCP-related activities but these were not fully operational. For example, the member of staff responsible for monitoring metal detection had not received full training.

At a company packing dry foods into vending sachets, the Auditor noted that there was no record to show that the relevant operative had received training for a particular CCP control – he was carrying out sieve checks.

At a tomato cannery, it was found that there was no evidence of training or experience for the seam analysis operative.

At a packer of oils and fats, the Auditor found that the (relatively new) QC operative was unsure of weight check and metal detection procedures and of the corrective action to be followed.

Nonconformities were given in each case.

Note: The clause also requires competency assessment of the trained staff.

It is not enough simply to train staff. How to assess the competence of staff involved in CCPs? This could be done in a number of ways: testing, supervision, information gathered from internal audits, complaints and HACCP reviews, to name some.

Clause 7.1.3

This is a requirement for documented training programmes and records. The Auditor will want to see documented programmes for training and these should include all aspects of training from induction through to refresher training.

Example

A cake bakery was found to have training procedures and records in place, but the procedures were not formally documented.
   A nonconformity was given.

The clause lists four key points that you must include in your programmes. Firstly, you must identify the competencies required for specific roles. You must then provide training to ensure that staff have the relevant competencies.

The review and audit of training is another requirement. The Auditor will expect to see evidence of review.

Example

A medium-sized speciality bakery had 120 staff and good initial training systems including induction and mentor-based training on the job. However, thereafter there was no system in place for monitoring staff effectiveness or reviewing training requirements.
   A nonconformity was given.

You may need to demonstrate the competency of the trainer. Finally, you must show consideration of training in the appropriate languages. As has been pointed out elsewhere, this last is now a very important aspect and the Auditor may want to see how you have trained staff on a multilingual site.

Clause 7.1.4

This covers the requirement for training records in detail and includes four key points that must be included as a minimum. They are self-explanatory. Remember to record the duration of training.

> ## Example
>
> At a company producing specialty breads, the duration of training was not in evidence for all staff training records.
>   A nonconformity was given.

The Auditor will also want to have a detailed look at general training records. For this, they are likely to pick out one or two members of staff at random, perhaps during the site tour, and ask to see their full records. Therefore, it is essential that your records are available for examination. Because this may be confidential information to you it is understandable that such files are usually secured under lock and key, so make sure that the key holder, such as your Personnel Manager, is available.

> ## Example
>
> A company producing tomato paste and passata was found to have no record of basic hygiene training for employees.
>   A nonconformity was given.

Note that you must also consider the situation where you have agency staff who have been trained by the agency. This is not unusual, for example agencies often supply staff to the food industry who are ready-trained in food hygiene. You must have records of their training available and the Auditor will certainly ask to see these. Effectively, you must have records available for all staff including agency workers.

Clause 7.1.5

This requirement concerns review. You must routinely review the competence of staff and where necessary modify the training procedures. There are a few ways to achieve this and the Auditor will look for evidence of staff appraisals and internal audits. The Auditor would also expect you to take account of product nonconformities and customer complaints in your review of staff training.

The Auditor will also expect you to keep staff training generally up to date. As legislation, customer requirements or other standards change (including the Global Standard), so training should be reviewed accordingly.

One option is refresher training, others are mentoring, coaching or simple time-served experience on the job.

Part Two

Clause 7.2: Personal Hygiene – Raw Material Handling, Preparation, Processing, Packing and Storage Areas

Personal hygiene is clearly of great importance in food manufacture. Your objective is that staff hygiene standards are product-related and documented. They must be seen to be understood and be acted on by all staff and visitors. Note that this includes agency staff.

Key points in Clause 7.2
Documented hygiene rules
Jewellery
Hand cleaning
Metal detectable plasters
Control of medicines

Clause 7.2.1

This requirement is about having documented hygiene rules that are communicated to all staff. The Auditor will study your rules to ensure that they are comprehensive and relevant to your site and that they include the five bullet points represented in this clause (note that this is a heavily merged clause comprising elements of five clauses from Issue 5).

You will need to have evidence available that you have trained staff in your rules. Preferably this should be done before they commence work, as part of their induction. Remember too that you might need to offer this important document in more than one language.

Examples

At a produce packer, there was no hygiene book or set of written rules for new workers.

A cheese manufacturer was found by the Auditor to have hygiene rules that were not a controlled document. Further, they were unclear with regard to procedures for control of personal medicine and staff using toilets while wearing protective clothing.

A snacks manufacturer was found to have documented hygiene rules referring to protective clothing and accessing toilets that were ambiguous.

A manufacturer of maize flour was found to have hygiene rules that lacked specific detail with regards to jewellery.

Nonconformities were given in each case.

The clause is very clear on what may not be worn in terms of watches and jewellery. *Note*: 'Sleeper earrings' are no longer permitted.

Examples

A speciality bread maker was found to have a jewellery policy but it was not clear what items were not allowed.

At a supplier of frozen pizzas, the Auditor noted that watches were inappropriately being worn by a member of the canteen staff and by staff in the stores.

A produce packer was found to have unacceptable jewellery being worn in process areas, for example hoop earrings, studs and watches.

Nonconformities were given in each case.

The reference to wedding bands means metal bands. Other kinds of bands are only permissible if they are not exposed at all. Jewellery must not be a potential contamination hazard. Where timekeeping is an essential part of the process, it may be permissible for an authorised member of staff to have a timing device in a production area; however, this should not be a watch worn on the wrist.

Concerning fingernails, the Auditor will look for signs of painted or false nails during the site tour and will also expect to see that this is reflected in your staff hygiene rules.

Example

At a poultry killing and processing plant, the Auditor found that there was no reference in the hygiene rules forbidding the use of false fingernails.
  A nonconformity was given.

This requirement also concerns perfume and aftershave. The requirement that it should not be excessive is simply because food can very easily pick up taints, especially from substances that are in effect designed for the purpose of changing the sensory profile. The Auditor will expect to see clear rules forbidding the wearing of strong perfumes and aftershave and will obviously be alert to this requirement during the site tour.

Example

At an ice cream factory, the Auditor found that there was no reference to the prohibition of excessive perfumes/aftershaves in the staff procedures.
  A nonconformity was given.

Part Two

Finally, you are required to check that your staff are actually following your rules. You might do spot checks or general audits but you will need to provide the Auditor with evidence of this.

Examples

An operative in the low-risk pre-pack area was noted wearing a hair slide beneath the mob hat, which contravenes the site hygiene rules.

At a tomato cannery, the Auditor found that employees were wearing watches, hairpins and jewellery and personnel were smoking, and some others were, without appropriate protective clothing, in the production area.

Appropriate nonconformities were given.

Clause 7.2.2

This is a specific requirement concerning the frequency of hand cleaning. The Auditor will expect to see adequate hand-washing facilities under Clause 4.8.6. Here, the Auditor will want to see instructions for hand cleaning on entry and at other times when necessary. This will depend on the circumstances and it is up to you to risk assess the situation to determine what is required. Commonly, such instructions are posted up at hand-wash stations, often using pictures. During the site tour, the Auditor will observe staff washing hands to confirm that they are following instruction. It is also something that you could include in internal audit.

Example

At a fresh poultry producer, the Auditor noted that some staff failed to wash hands at entrance to raw processing area.
    A nonconformity was given.

Clause 7.2.3

This is a very straightforward requirement that needs no explanation. All wounds must be covered by appropriately coloured plasters, preferably blue and they must contain a metal detectable strip. The idea of blue, metal detectable plasters is well known throughout the food industry and they must be used whether you are metal detecting or not. They shall be issued by the company in a controlled way and staff must not use their own dressings.

In the past, it was not unusual for companies in certain countries outside the United Kingdom to claim that blue metal detectable plasters were not available in their country. I am afraid that this excuse will no longer be tolerated now that such plasters are available through the Internet. Try typing 'blue metal detectable plasters' into your favourite search engine and you will be greeted with numerous suppliers.

All staff should be made aware of this requirement and the Auditor will want to see a system for issuing plasters and will look to see that any plasters in use meet the requirement.

Examples

At a producer of fresh tomatoes, the Auditor found that not all plasters being worn were metal detectable.

A manufacturer of egg products was found to have no reference to the compulsory use of blue, metal detectable plasters in the documented personal hygiene rules.

A nonconformity was given in each case.

Where you have to further cover a plaster, you must use a glove. The clause does not specify the type of glove but clearly it must be a food-safe item such as used for normal food handling.

Clause 7.2.4

Leading on from the previous clause, this requirement is also very straightforward. If you have metal detection, then you must check your blue metal detectable plasters through at least one of them. The Auditor will want to see records as evidence that you do this with some regularity. It is required that every batch of plasters be tested (some companies record this with a sticker on the box). The Auditor may ask you to check one during the audit as part of the tour.

Examples

At the time of an Audit of a processed cheese manufacturer, they had only recently installed a new metal detector. The Auditor found that they had an informal system for checking plasters and that this was not adequate.

Records were not being kept at a bakery to demonstrate that a sample of each batch of blue plasters was being tested through the metal detectors.

At a dry ingredients manufacturer, the Auditor found that there was no evidence that the current batch of blue plasters had been tested through the metal detector.

A nonconformity was given in each case.

Part Two

Some suppliers have expressed concern over this requirement where circumstances make a meaningful test difficult. For example, if you are packing a large item such as a full block of cheese or a catering pack of something, then the sensitivity of your metal detector may not be high enough to detect the plasters. This issue has certainly been discussed among UK-based CBs and you can expect the Auditor to apply common sense in such circumstances. However, this should not be an excuse where the Auditor can see that your metal detector is insufficiently sensitive for your products.

Clause 7.2.5

This requirement concerns the control and storage of personal medicines. The idea is that they could become a hazard to the product if not properly controlled. The Auditor will expect to see procedures in the form of work instruction or perhaps as part of staff hygiene rules. It is expected that staff should not generally be permitted to take medicines into production areas. Where persons have specific conditions that might require urgent taking of medicine (e.g. asthmatics, angina sufferers, diabetics and so on), you should make special provision so that the medicines are available to them in a controlled way. The Auditor will take a pragmatic view of this but you must demonstrate that medicines are controlled.

Example

A fruit juice manufacturer had a serious problem because of a syringe found in one of their packs. I had to visit the site during this period and was told that the company had been aware of a potential problem before the incident when a number of diabetic syringes had been found on the floor of a storage area.

In this case, the company acted very properly and openly after the event and, of course, made some significant changes to their control of such items for the future.

Other Examples

A bottler of mineral waters was found to have no personal medicines policy.

In a pasta factory, the Auditor found that in the hygiene rules there was no reference to the control of personal medicines.

Nonconformities were given in each case.

Remember too that although the clause specifically refers to staff, common sense dictates that there should be control over visitor's personal medicines also.

Examples

A ready meals manufacturer was found to have a system of reporting personal medicines under the staff hygiene rules, but this did not include visitors to the site.

At a cheese factory, it was found that the personal medicine control for visitors to the site was verbal and unstructured.

Nonconformities were given in both cases.

Clause 7.3: Medical Screening

This SOI concerns all employees (including agency staff) and visitors where product safety is concerned. Essentially, it is for you to set a policy on what level of medical screening is appropriate and it will depend on the nature of your products.

> **Key points in Clause 7.3**
> Preventing transmission of food borne disease
> Questionnaires
> What to do if you are aware of infectious disease

In many cases, declarations from personnel by medical questionnaires will suffice, while in particular high-risk operations it may be necessary to go as far as stool testing. In some countries, there are legal requirements concerning the medical standards of workers in the food industry. Some countries also have privacy laws relating to such records that can make audit more difficult. In other cases, the employer can have a certificate issued by the relevant government authority confirming suitability for working in the food industry.

The Auditor will want to see a system in place that covers all relevant staff and visitors and some records to confirm this.

Part Two

Examples

At a manufacturer of grated and processed cheese, the Auditor found that there was no evidence that all staff have completed a medical questionnaire prior to initial employment.

At a produce supplier, medical screening questionnaires for staff were not available to the Auditor for examination.

Nonconformities were given in both cases.

Clause 7.3.1

Personnel must be made aware that should they contract or be in contact with a relevant infectious disease such as a food borne infection or food poisoning, they must notify the management. Note that this requirement includes all temporary workers.

Example

At a cake bakery, it was noted that the return to work questionnaire concerned personal health and safety rather than food safety.
    A nonconformity was given.

Clause 7.3.2

This is a relatively straightforward requirement concerning visitors and contractors. Where they are going to enter food areas or for any other reason that may affect food safety, they must complete a health questionnaire or indicate in some way that they are not suffering from a condition that puts food safety at risk. This should include all such visitors, so, for example remember to include your pest control contractor and, of course, your Auditor. The Auditor will ask to see evidence of some completed ones (other than their own) such as your pest controller. The Auditor may have noted when the previous pest control visit was made and check this against your entry system.

Example

At a produce packer, the Auditor noted that the pest controller had not been asked to complete a visitor's questionnaire when they last entered the site.
    A nonconformity was given.

Clause 7.3.3

This requirement concerns your action in the event that you become aware of any person suffering from a relevant disease who may enter the premises in that condition. The Auditor will want to see rules in place for this (i.e. documented procedures) and will also question what action you take in such circumstances. It will be expected that the rules prevent an affected worker or visitor from entering food areas until they are clear of the disease. This might require clearance from a doctor. The Auditor will want to see a system here, which ensures that the person concerned follows the procedure and that you take whatever action that would be needed to ensure product safety. This must be a written procedure, which is communicated to the staff.

Thus, if you are aware that the person has come into contact with product and may have affected its safety, you should have an action plan for possible isolation and quarantine of affected product. Such product would then be treated under your nonconforming product systems.

Clause 7.4: Protective Clothing – Employees or Visitors to Production Areas

This is a very straightforward SOI that anyone entering food handling areas shall wear suitable protective clothing that is issued by the company.

> **Key points in Clause 7.4**
> Documented rules on attire
> Having sufficient clothing
> Covering hair
> Laundry and laundry standards
> High-care/ high-risk laundry
> Gloves
> What to do about items that cannot be laundered

The Auditor will expect you to have sufficient clothing for all staff and visitors, including themselves. The suitability of the clothing will vary depending on the nature of the area and type of work. For example, a higher level of protection will be expected in a high-risk food area. More detail on suitability is covered in the subsequent clauses.

All visitors should be given clothing by you if they are entering food handling areas. Local authority officials such as environmental health officers in the United Kingdom are sometimes fond of turning up with their own overalls. Do not allow this: you do not know where they have been! There is a risk analysis element to this though. You might make an exception with your pest controller, for example as their overalls are usually very distinctive and they are usually well away from exposed product.

Note: The SOI and all subsequent clauses refer to production areas only.

Clause 7.4.1

You must document all your rules concerning the wearing of protective clothing and you must communicate this to all. This means that even for visitors and contractors you have to make sure they have read your rules on clothing. Clearly, your Auditor will expect to have seen these rules before entering the factory, but may also ask for evidence that other visitors have seen them.

As far as staff go, this is likely to be a part of your normal training, probably induction. The Auditor will want to see evidence that all staff have been made aware of these rules and, just as with the general rules on personal hygiene, may pick a few staff names out and ask to see the relevant training records.

Note that the rules must include some specific measures such as any necessary changing between low and high risk, and any requirement to remove items before visiting toilets or smoking areas. The Auditor will note how such rules are observed during the site tour.

Examples

At a creamery, the evaluator found that some members of staff did not observe the overalls procedure for visiting toilets (unclear instructions in hygiene rules).

Protective clothing was worn outside against company procedures.

In a dairy producing farmhouse cheese, it was found that high-care operatives' clothing was not removed prior to entering toilets.

A nonconformity was given for each case.

In some cases, you may have to carry out a risk assessment, for example where staff wear company-issued tops and trousers (e.g. 'baker's whites'). Should they remove trousers?

Clause 7.4.2

This requirement firstly prescribes that you have sufficient clothing for all employees. The Auditor will be aware of your staff numbers and will also ask how many changes of clothing your staff are required to have in a week.

Protective clothing shall be designed to ensure product safety. The clause goes on to specify the style of protective clothing required in that it shall have no external pockets (at least above the waist) or sewn-on buttons. They must also be in good repair and clean. In terms of overalls, the Auditor will be looking for coats and aprons that properly cover personal clothing. If the industry dictates that all personal clothing is changed, as in the bakery industry, where baker's whites are the norm, then this requirement is satisfied in that way.

Examples

In a manufacturer producing dry food ingredients, the Auditor noted that buttons were used on trousers worn in the factory.

A number of pack house operatives were noted wearing protective clothing with external pockets.

At a poultry processor, it was noted that many of the staff were wearing inappropriate protective coats with external top pockets.

A nonconformity was given in each case.

There is also a requirement concerning hair and head coverings. Essentially all hair should be covered, that is scalp hair. The Auditor will look at conditions of headgear during the site tour. Hair must be covered in all parts of the site where product is exposed. There may be other areas, such as a warehouse, where such a regime is not necessary. As always, it is for you to risk assess the situation.

Examples

At a pulses cannery, the Auditor found an instance of hair not fully contained by the headgear.

In a tomato cannery, the Auditor noted that there were instances of operatives in production, particularly maintenance, not wearing head covering.

A nonconformity was given in each case.

Finally, there is what appears to be a straightforward requirement but is one that can cause debate. As far as facial hair is concerned, you must consider it in the same way as hair generally and that it should be covered where there is potential risk to product. Then, there is the thorny problem of what exactly is a beard or moustache. I can only express an opinion on this. In my view, if a member of staff has significant facial hair and will be working with exposed food then he is a risk to the product whether it can be called a beard or moustache or sideburns or whatever. For example, a couple of day's growth of stubble is not usually considered a problem but a long handlebar moustache would be.

The Auditor will expect to see a beard snood policy that staff are aware of. Where you have staff with facial hair, they must be wearing snoods where product is at risk. As with head coverings, in areas where food is not exposed, such as warehouses, snoods may not be required, depending on your risk assessment.

Part Two

Examples

I was once at a fresh meat cutting plant where the audit was not going terribly well. It did not help matters that the consultant they had brought in to help had a heavy beard and was not offered a snood for the site tour, nor did he ask for one.

This resulted in one of many nonconformities.

In a processed cheese manufacturer, the Auditor found that there were no beard snoods available while a bearded employee was present in the processing area.

A nonconformity was given.

Clause 7.4.3

This clause concerns the laundering of such protective clothing. Clothing may be either laundered in-house or by an approved laundry, which must also be under contract as with any service provider (see Clause 3.5.3.2, Chapter 13). Furthermore, the effectiveness of cleaning must be monitored. If you are laundering in-house, the Auditor will expect you to show evidence of proper procedures and control systems. The clause requires that you have defined and verified criteria to validate the process.

As far as monitoring is concerned, you must show some method for this, based on detailed procedures. Depending upon the industry, some sites will have visual checks, while occasional swab tests are now seen in high-risk plants.

Examples

A manufacturer of chilled and frozen ready meals was laundering in-house. It was found that a documented laundry procedure (in-house) and verification and control system for it were not available.

A grower and packer of onions, shallots and garlic had an in-house laundry system; however, the Auditor found that there was no formal procedure for this.

A nonconformity was given in each case.

Where you use an outside contractor, the Auditor will want to see evidence of supplier approval as for any contractor (see Clause 3.5.3.1) and some checking on the cleanliness of the clothing.

Examples

The Auditor at a poultry slaughtering and cutting plant found that the company was using a laundry contractor. However, the effectiveness of laundering processes was not monitored.

The Auditor at a bakery found that detailed procedures to ensure the effectiveness of the laundering process were not in evidence.

A nonconformity was given in each case.

The washing of workwear by your staff, in other words, taking it home, may only be done in exceptional circumstances and only where:

- the clothing is to protect the employee from the product and not the other way round; and
- the clothing is for enclosed product or low-risk areas only.

Note: On no account may high-risk or high-care overalls be laundered by the staff themselves.

Example

A poor situation was noted at a bakery where the documented laundry procedure described actions to be taken when workwear was laundered in-house, that is on site; however, laundry was now cleaned by staff at their own homes with no proper procedures.
   A nonconformity was given.

In certain industries such as harvesting on produce rigs, there is a practice of staff washing their own overalls at home. In general, the washing of overalls by staff at home is not encouraged but is not specifically forbidden by this requirement.

Clause 7.4.4

This is a new clause for Issue 6 concerning laundry of clothing for high-care and high-risk areas. In such cases where you are using a contractor, the contractor must be audited. This may be done by yourselves or by a third party.

The clause lists four bullet points that your laundry must satisfy. The Auditor will want to see evidence of audit or at least a certificate and some evidence that they meet the four bullet points. Thus, you may have to provide a copy of the contractors procedures, although the Auditor might see evidence during the site tour of some aspects, for example protection of cleaned clothes.

Example

Cleaned workwear for high-care cutting room area was transferred unprotected through the main warehouse before entering the changing facility.
   A nonconformity was given.

You may also be able to verify the sterility of clothing yourselves after washing by swabbing. Some companies do this. You can certainly keep a check on how the laundry delivers the items to your site and whether they are protected in bags, for example.

Clause 7.4.5

This requirement concerns the control of any gloves issued. They shall be controlled to avoid product contamination.

There are several types of gloves that are encountered by the Auditor. Where food is directly handled, vinyl gloves are the norm. The Auditor will expect to see that they are in good condition and that there is some control over their issue and disposal. Gloves should not be allowed to become tattered or dirty and a hazard in themselves. The clause requires that they are replaced regularly. If appropriate, they should be disposable. It is also required to have them in a non-food colour such as blue.

Examples

A chilled hamburger meats and sausages manufacturer was found to be using gloves that were flesh coloured.

It was noted during a factory inspection, that one member of staff was using clear sellotape to protect fingertips from rubbing.

At a soft drinks supplier, it was found that there was no procedure in place to ensure that gloves worn did not contaminate product.

At a poultry processor, it was found that control of issuing of green gloves was not being carried out as per procedure.

At a food ingredients manufacturer, it was found that the gloves issued were subject to no controls.

Nonconformities were given in each case.

Note: The use of latex gloves should be avoided because of the risk of allergic reaction. This could affect the staff or even a consumer.

Other gloves such as oven gloves should be kept in good condition and the Auditor will look at these during the site tour to ensure that they are not a hazard to product integrity.

Clause 7.4.6

The final clause! It is a new one for Issue 6 concerning items that cannot be laundered such as chainmail items. You might also consider high-visibility ('hi-vis') waistcoats and the like that are sometimes seen in production areas. The Auditor will ask how such items are cleaned and sanitised and at what frequency. For this one, you might need to carry out a risk assessment.

Summary

This part of the Standard has changed much over the years as awareness of the importance of training and other staff matters has grown. Good training underpins everything that you set out to do and your relationship with staff is so important. Staff must also be given the right environment for the job, it is important for food safety and for staff morale. In many cases, they are your biggest investment and rightly so. But remember also that standards for visitors and contractors are also important.

Quiz No. 12

(1) Are staff permitted to wear any earrings?
(2) At what rate should you test metal detectable plasters?
(3) Overalls may have no external pockets, true or false?
(4) Who is responsible for keeping records of agency-trained staff?
(5) How must you approve the laundry for high-care/high-risk production?

Part Three
After the Audit

In this part, I look at the events immediately after your site audit, leading up to certification. Then, I discuss your continued certification until your re-audit is due.

The chapters are:

18 From Audit to Certification

We now return to the Protocol and what it says about events after your audit. In this chapter, we will look more of the paragraphs of Part 1 of the Protocol (we have already discussed the essential points of Part 2 of the Protocol in Chapter 5 as they are concerned with audit options and therefore relate to events before the audit).

Key points in this chapter
Corrective action and root cause
Two important timescales
Differences for critical, major and minor nonconformities
Technical review by the CB
The report
The certificate

Paragraph 9.2: Procedures for Handling Nonconformities and Corrective Action

Paragraph 9.2 of the Protocol is concerned with the subject of handling nonconformities raised during the audit and corrective action and root cause analysis.

If you have nonconformities raised, they must be corrected by you within the stated timescale (see Section 'Timescales'). The correction of nonconformities is considered to be a part of the audit process; in other words, the audit is not over until this part of the process is completed.

Note: The following applies when you have had a normal audit. If you are in the Enrolment Programme or continual development scheme and therefore do not aspire to a certificate this time, you need only submit corrective action plans (see Chapter 5).

Root cause

Here is something new, introduced for Issue 6. We discussed this and touched on some of the methods available in Chapter 4. For any nonconformity, as well as correcting

The BRC Global Standard for Food Safety: A Guide to a Successful Audit, Second Edition. Ron Kill
© 2012 John Wiley & Sons, Ltd. Published 2012 by John Wiley & Sons, Ltd.

Part Three

the immediate issue, you must now identify the root cause of the issue. What does this mean? Any issue must have an underlying cause, whether it is something to do with resources, training, hazard analysis, factory layout or whatever, you must identify this. Furthermore, when you submit your corrective action for each nonconformity, you must also submit an action plan to address the root cause. This plan will feature in the audit report alongside the corrective action.

> ## Try This
>
> Imagine you had been given the seven nonconformities in Table 5.4. Jot down what you think would need to be the root cause action plan in each case.

Timescales

The timescales for completing the audit are more important than ever. Overall, the CBs must complete the entire exercise in 49 days (see Section '49 days').

I must stress the importance of the timescales in submitting corrective actions. If you fail to meet the timescales, then it is likely that you will not achieve a certificate. This is also an aspect on which the CB's performance is measured by the BRC, so they will be chasing you hard to ensure you get corrective actions in on time (see Section 'Choosing a Certification Body', Chapter 7).

28 days

There is a strict deadline to be observed. For all nonconformities, your corrective actions must be with the CB no more than 28 days after the audit. The CB will enforce this strictly and may send you reminders if nothing has been received and the 28th day is approaching. However, the obligation is on you to send in whatever material is necessary and there is no obligation on the part of the CB to keep reminding you of this.

Regarding the 28-day period, the CB may use discretion in the case of exceptional circumstances.

> ## Example
>
> I was involved with a company that had a murder take place on its premises shortly after an audit. As you can imagine, this was an unusual enough circumstance to allow a little leeway on their deadline.

The CB is under pressure itself to complete the audit within the timescale and will not allow unwarranted delays. So, please remember to submit all your corrective actions within this deadline and allow for delivery times if using postal services. Most companies are sending corrective actions in by email and this can certainly help (see Chapter 19). During the technical review (see Section 'Technical review and certification decision'), the CB may ask you for more information, so your timekeeping on the initial submission is very important – the earlier the better. If you can get corrective actions in sooner than the 28 days, please do so.

In the case of a Grade C being awarded, the CB must return to site to verify corrective actions within 28 days. It is likely that a date for this will be arranged at the closing meeting or shortly after.

Note: If there is no commitment from you to complete the corrective actions within 28 days without any mitigating reason, then you will have to go through a complete re-audit in order to achieve certification.

49 days

The CBs are obligated to the BRC to complete the audit, reporting and certification processes within 49 days of the audit date. Thus, unless you have critical noncon-formities or major nonconformities against Fundamental Requirements and assuming your corrective actions arrive within 28 days *and* they are satisfactory to the CB, the CB has effectively a maximum of 3 weeks to assess the audit, assess the corrective actions and present the material to a certification committee and arrive at a certification decision (see Section 'Technical review and certification decision').

Note that officially the CBs were originally given 42 days to complete the job and many still work to this deadline to give a little leeway for contingencies. Indeed, the Standard still requires reports to be 'dispatched to the company within 42 days'.

Uncertificated

Where the result of the audit is 'Uncertificated', that is 'No Grade', there is no set timescale, because you will effectively have no current certificate and will need a complete re-audit at a date of your choice. It is considered unlikely you would be able to fully correct such a position within the usual 28-day period.

Note: Any previous certificate still valid will be withdrawn.

Nature of corrective actions

The CB must be able to verify all corrective actions. Corrective action must not only be definitive, it must be supported by evidence. In many cases, this can be done with documentary evidence, for example records, photographs and procedures. Modern technology is also increasing the scope of supplying evidence; for example, recently we have received video clips by email of certain features actively working, which was very useful. However, aside from a Grade C situation, the CB may also need to return to site to verify corrective actions especially where physical evidence is not easily conveyed by other means. For example, in case of structural changes that are not possible to verify from a distance.

Part Three

In Chapter 19, I return to this subject with more detail on how to present corrective actions.

Differences in correcting critical, major and minor nonconformities

Critical

For any critical nonconformity, the site will be uncertificated (given 'No Grade') and a complete re-audit will be necessary. As we discussed in Chapter 2: Failure, there are other circumstances where you might be uncertificated.

Note: For any critical nonconformity or otherwise failure to gain certification, your customers may require you to inform them of the situation.

Major and minor

For a major nonconformity against a Fundamental Requirement, a site will be uncertificated. For any other major or minor nonconformity, you must submit corrective action within the stated timescale. This must be definitive and completely close out the nonconformity. Temporary solutions are permitted provided they are acceptable to the CB. The CB must be able to verify corrective action and this may mean a return to site. Remember too that you must undertake a root cause review and submit this action plan in the same timescale.

Note: If the Auditor cannot verify the effective close-out of nonconformities at the subsequent audit, or if the root cause has not been resolved, you will get a nonconformity against Clause 1.1.10.

Paragraph 10: Grading of the Audit

The complexities of grading and consequent audit frequencies are well presented in the table in the Protocol relating to this paragraph and need no detailed explanation here. Essentially, it is all down to the number and nature of nonconformities raised at your audit. The grading system was thoroughly revised for Issue 5 and remains effectively the same for Issue 6 except that some anomalies have been corrected and uncertificated status is no longer described as Grade D but as 'No Grade'. Note that where a CB gives 21 or more minor nonconformities yet no majors, they have to justify this in the report (i.e. explain why no majors were given).

Technical review and certification decision

At the end of the audit, you will have a good idea of your grade, and you can do the sums yourself from the table. However, the grading process does not happen at the site but only after technical review by the CB looking at all the circumstances of the audit. A technical or certification committee will go through the draft audit report, including any nonconformities raised and your corrective actions and root cause action plans. It is possible that this review might lead to changes in the number and/or nature of nonconformities and, of course, the CB will inform you of this before issuing a certificate. It is also possible that the CB will find your evidence unclear or incomplete and ask you for more information or evidence. It is important that the communication channels between the companies are good therefore and that both sides respond as quickly as possible. See Chapter 19 for more on this process.

Paragraph 11: Audit Reporting

The report can only be finalised and issued after the certification decision because it must include that decision and brief details on the technical review of corrective action evidence and root cause plan. For Issue 6, the format of the report was changed. The front sections of the report have been expanded to include more detail on the site, such as employee number, site size and so on. For those familiar with previous issues, the new report no longer requires detail on the evidence seen for each single clause. For Issue 6, the final part of the report gives an overview of performance against each main clause of the Standard and a simple indication of conformity against the individual clauses.

Note: As before, any language may be used; but if the summary sections are reported in a language other than English, then they must be reported in English as well.

There are two formats for the report: one for the standard audit and one for an Enrolment audit where the company is in the continual development phase. They do not differ greatly apart from for the Enrolment version: the inclusion of the scores in each section, no column for evidence for the nonconformity section, plus, of course, they have different colour schemes.

An important matter is the ownership and availability of audit reports. There is no doubt in the minds of CBs that it is the company being audited that owns the report: you are the client and you are paying for it. It is considered to be confidential by the CB, and a commitment to confidentiality should be a part of the contract between the parties. However, the CBs are also obligated to upload a copy to the BRC via the online Directory (see Chapter 8). Note that the CBs are required to upload reports as soon as possible even if the certificate has not been actually issued to you yet. Many CBs will delay issuing certificates until payment has been received. Once uploaded, the idea is that you may then give safe access to whichever customers you please to your own reports.

Paragraph 12: Certification

Paragraph 12 refers to the certification process and format. The certificate has to be in a certain format and contain certain information that is set out in Appendix 5 of the Standard. An example is also given in Chapter 20. Some of the detail required is also discussed and expanded on here. One point for you to note is that when you receive your certificate, you must check the audit scope that has been put in by your CB to make sure it is stated correctly for you. If there are any problems with it, you should discuss it with your CB and, if necessary, get them to reissue it.

Note: Unlike the report, the certificate remains the property of the CB.

Part Three

Summary

In this chapter, we have looked at some very important processes that involve both you and the CB. The period immediately after the site audit is critical in terms of the communication between the two parties and in terms of timescales. See Chapter 19 for some detail on following up the audit.

Quiz No. 13

(1) What is the deadline for corrective actions?
(2) Who owns the report?
(3) Who owns the certificate?
(4) When is the audit grade finalised?

19 Correcting Nonconformities

Well done, you have got through the on-site audit. However, unless you are in the continual development phase, the audit is not complete until all nonconformities have been satisfactorily corrected. Unless you are a rare species with no nonconformities, you must go through a corrective action process.

Key points in this chapter
Stating corrective action clearly
Making sure the evidence is comprehensive
Types of evidence
Follow ups
Returning to site
Root cause action plans

In Chapter 5, I discussed nonconformities in all their possible permutations and in Chapter 18 indicated the essential protocols for corrective actions. If you have been given no grade, then a complete re-audit is necessary with no timescale imposed. In the case of a C grade, you must have your corrective actions verified on site within 28 days. In all other cases of nonconformity, it is necessary to submit corrective actions with supporting evidence within 28 days of the audit to the CB along with root cause action plans.

This chapter describes how best to present your corrective actions and root cause plans. We will look at corrective actions first.

Note: The following details apply to an audit of a certificated company. For the Enrolment Programme different conditions apply (see this chapter, Section 'Enrolment Programme').

Documentary Evidence

In many cases, the CB will be able to verify corrective actions without returning to your site (see this chapter, Section 'Returning to Your Site'). You must understand the need for keeping to a tight deadline and for supplying the CB with *evidence* of

The BRC Global Standard for Food Safety: A Guide to a Successful Audit, Second Edition. Ron Kill
© 2012 John Wiley & Sons, Ltd. Published 2012 by John Wiley & Sons, Ltd.

your corrective actions. It is in your interest to get this right the first time. If the CB is not satisfied with your submission, they will then have to ask you to submit further evidence. The clock is always ticking and this may take you over the permitted time. If this process takes too long, you may not achieve a certificate. During this period, you need to be as helpful as possible to the CB.

Stating the Corrective Action

Essentially, when sending in your corrective actions, you should start with clear statements of what action you have taken. You should not just send in a batch of documents or other evidence with no explanation. The CB is likely to have left you with a form to complete for each corrective action to encourage you to do this. If so, you should use the form because it will assist the CB greatly. Speaking from personal experience, it is very difficult to assess properly the validity of corrective actions if the company simply sends in just the evidence of corrective action with no information as to its relevance. It is worse still if there is 'evidence' but no reference to the nonconformity clauses concerned. If the CB has not supplied you with a corrective action form, then put down your actions in your own format to accompany the evidence. These should be clear and concise and refer to each nonconformity.

The following is an example of a company submitting clear statements of corrective action and acceptable documentary evidence.

Example

A seafood supplier had been given one major and six minor nonconformities. These are shown in Table 19.1.

The company submitted properly completed corrective action forms to the CB accompanied by evidence as presented in the following pages.

Table 19.1 Nonconformities given during audit at sea food supplier.

No.	Requirement ref.	Detail of nonconformity
Major		
1	1.1.4	There was no programme for meetings for food safety and legality issues.
Minor		
2	3.4.1	The internal audit schedule was unclear and HACCP audit scheduled for May had not been conducted.
3	3.5.1.3	The supplier approval 'exceptions' procedure had not been clearly defined with reference to agents.
4	3.5.3.1	They had a supplier of services approval and monitoring and had approved most suppliers in the procedure but had not included their occasional sub-contractor for transport.
5	4.4.5	There was mould growth on the ceiling in the fish preparation area.
6	4.5.4	There was no record of the compressed air filtration 6 monthly change (last done xx/xx/xx)
7	7.3.2	The pest control biologist did not sign on to site or complete a health questionnaire on xx/xx/xx.

BRC corrective action sheets and comments follow in pages 332–345.

Part Three

BRC Audit Corrective Action Sheet

Supplier	Sea Food Ltd	Audit date	xx/xx/xx
Site address	Fish St, Grimsby, UK		

Nonconformity reference	Details of corrective action taken	Details of evidence supplied
1.1.4	Monthly meeting programme created to discuss issues and first meeting held.	Programme for 12 months. Minutes of first meeting.

NAME .... D. Finn ....

SIGNATURE .... *D Finn* .... DATE .... xx/xx/xx ....

Comment on corrective action and evidence supplied for Clause 1.1.4

This was a major nonconformity against this clause because they had completely failed to meet the requirement.

In their corrective action statement, the company clearly stated that they had written a programme for scheduled meetings and supplied a copy of this programme as evidence. The programme was for the next 12 months.

In their corrective action statement, they also stated that they had held their first meeting. Evidence of this was supplied and the CB felt that considering the evidence was supplied within 28 days there would be no more that the company could have done in this timescale.

Therefore, it was reasonable to close out this nonconformity at this time. Of course, their continued adherence to this schedule would be checked at the next audit.

BRC Audit Corrective Action Sheet

Supplier	Sea Food Ltd	Audit date		xx/xx/xx
Site address	Fish St, Grimsby, UK			
Nonconformity reference	Details of corrective action taken		Details of evidence supplied	
3.4.1	Schedule updated, HACCP audit carried out and completed		New audit schedule and HACCP audit report	

NAME . . . . D. Finn . . . .

SIGNATURE . . . . *D Finn* . . . . DATE . . . . xx/xx/xx . . . .

Part Three

Comment on corrective action and evidence supplied for Clause 3.4.1

This was a minor nonconformity because although their schedule was unclear they had completed all their required audits apart from the HACCP audit, which was delayed.

The company have clearly stated what they have done and in support of this have supplied a detailed revised audit schedule in tabular form. It had been updated to show all the audits carried out during that year, including the one carried out on the HACCP system since the audit.

They also submitted a report for the audit carried out on the HACCP system. This report clearly showed items that compiled with their procedures as well as some internal nonconformities. The report was dated and signed.

The CB considered this evidence sufficient. The company had completely corrected this nonconformity by improving the schedule and carrying out the audit that the Auditor had found to be missing and they had supplied evidence of it.

Thus, the CB was able to close out this nonconformity.

BRC Audit Corrective Action Sheet

Supplier	Sea Food Ltd	Audit date	xx/xx/xx
Site address	Fish St, Grimsby, UK		
Nonconformity reference	Details of corrective action taken		Details of evidence supplied
3.5.1.3	A revised procedure for supplier approval has now been written and 'Exceptions' has been documented as step 7 of the procedure to include the use of agents.		Supplier approval procedure ref: FSQMS/2.10/3

NAME .... D. Finn ....

SIGNATURE .... *D Finn* .... DATE .... xx/xx/xx ....

Part Three

Comment on corrective action and evidence supplied for Clause 3.5.1.3

This was a minor nonconformity against this clause because they had a procedure in place but it was not comprehensive, missing the reference to the use of agency supply.

This was a straightforward nonconformity, which was straightforward to correct. Thus, the corrective action was clear in that they had rewritten their supplier approval procedure, which included a new section concerning the use of agents that supply minor ingredients.

The company procedures were now complete and thus now met the requirement.

BRC Audit Corrective Action Sheet

Supplier	Sea Food Ltd	Audit date		xx/xx/xx
Site address	Fish St, Grimsby, UK			
Nonconformity reference	Details of corrective action taken		Details of evidence supplied	
3.5.3.1	Sub-contractor completed supplier questionnaire.		Copy of completed questionnaire.	

NAME .... D. Finn ....

SIGNATURE .... *D Finn* .... DATE .... xx/xx/xx ....

Comment on corrective action and evidence supplied for Clause 3.5.3.1

This was a minor nonconformity against this clause because they had a procedure in place but it was not comprehensive in missing out one supplier of a service.

In their corrective action, they got the sub-contractor to complete a supplier questionnaire and the evidence showed this and that the company had checked and approved it.

The CB was able to close out this nonconformity.

BRC Audit Corrective Action Sheet

Supplier	Sea Food Ltd	Audit date	xx/xx/xx
Site address	Fish St, Grimsby, UK		
Nonconformity reference	Details of corrective action taken		Details of evidence supplied
4.4.5	The ceiling was cleaned and the staff were reminded to check this during the weekly hygiene inspections.		(1) Photographs of ceiling before and after cleaning (2) Memo to staff to include in weekly hygiene audit (3) Copy of latest hygiene audit

NAME .... D. Finn ....

SIGNATURE .... *D Finn* .... DATE .... xx/xx/xx ....

Part Three

Comment on corrective action and evidence supplied for Clause 4.4.5

This was a minor nonconformity against this clause because only a part of one area of the ceiling was affected.

Their corrective actions here are quite comprehensive and they have anticipated that just sending in photographs of the cleaned ceiling is not sufficient because this is a situation that could easily arise again. Therefore, they have ensured that they monitor the situation better and have sent in good evidence of this.

The CB was able to close out this nonconformity.

BRC Audit Corrective Action Sheet

Supplier	Sea Food Ltd	Audit date	xx/xx/xx
Site address	Fish St, Grimsby, UK		
Nonconformity reference	Details of corrective action taken		Details of evidence supplied
4.5.4	Record was not seen on day of audit as documentary evidence was not stored in correct place. Compressed air filtration was serviced and filters replaced on xx/xx/xx		Service record and schedule attached

NAME .... D. Finn ....

SIGNATURE .... *D Finn* .... DATE .... xx/xx/xx ....

Comment on corrective action and evidence supplied for Clause 4.5.4

This was a minor nonconformity against this clause.

The nonconformity concerned a record not being present. The corrective action stated that in fact the record existed but had not been in the right place at the time of the audit. As evidence for this, they submitted the missing document, a service record for the compressed air filters dated before the audit date. They also submitted a procedure for the service itself provided by the contractor.

This effectively closed this nonconformity.

BRC Audit Corrective Action Sheet

Supplier	Sea Food Ltd	Audit date		xx/xx/xx
Site address	Fish St, Grimsby, UK			
Nonconformity reference	Details of corrective action taken		Details of evidence supplied	
7.3.2	Pest control biologist informed of company procedure in writing. Copy of medical screening form and permit to work placed in pest control log book		Letter to contractor	

NAME .... D. Finn ....

SIGNATURE .... *D Finn* .... DATE .... xx/xx/xx ....

Comment on corrective action and evidence supplied for Clause 7.3.2

This was a minor nonconformity against this clause.

The company had clearly stated their corrective action in that they had written to the pest control contractor. The letter supplied as evidence explained that the site had a procedure for all visitors and contractors to sign in and that while there was a shared responsibility for this, they requested that the pest controller's staff be made aware of this procedure.

The company also stated that the signing-in form had been placed in the pest control log as a reminder although, clearly, they could not reasonably supply evidence of this.

As the company had done as much as it reasonably could, given that the pest controller was not scheduled to visit the site during the 28-day corrective action period, the CB was happy with the corrective action. The continued compliance would be checked at the next audit.

Conclusion

The CB was satisfied that all the requirements were now met with no further information required.

Provided the company also supplied acceptable root cause action plans (see this chapter, Section 'Root Cause') they will have achieved a Grade B certificate.

When this method is not followed, it can be difficult for the CB to be clear about whether any evidence supplied is corrective and what your intentions are. In the following example (see Table 19.2), no corrective action statements were supplied.

Example

A manufacturer had only three minor nonconformities to correct as seen in Table 19.2.

Some days after the audit, a bundle of papers was received by the CB with a covering letter but no detailed statements on the corrective actions taken. Sifting through the evidence presented that there were some packaging specifications and product specifications that were not dated or signed. There was a record sheet showing checks on scales.

Thus, after a little deliberation, the CB was able to close out Clause 6.3.2 as the evidence did point to what they had done and they now met requirements. However, it was necessary to seek further information and clarification for Clauses 3.6.3 and 3.8.1. For the latter, there was no clear corrective action at all.

This clarification was received in time, thus in this case, the company was able to satisfactorily close out the nonconformities properly within an acceptable period, but significant time and effort was wasted in what should have been a simple matter of correcting only three minor nonconformities.

Table 19.2 Example of three minor nonconformities.

No.	Requirement ref.	Detail of nonconformity
1	3.6.3	The current finished product specifications were incomplete and did not include detail on legislation, nor were they authorised.
2	3.8.1	Handwritten records on product nonconformance were not retained.
3	6.3.2	In-house scale calibration checks not recorded.

The Nature of the Evidence

Assuming that the CB will not need to return to site to verify corrective action (see this chapter, Section 'Returning to Your Site') evidence supplied remotely to the CB must be definitive, whether it be for a major or minor nonconformity. For example, suppose the nonconformity concerns a procedure not having been followed and recorded. Your corrective action statement might be:

> New form designed to record routine cleaning and relevant staff retrained to complete the form.

However, if you accompany this with a blank example of the form as your evidence, this will not satisfy the CB. Indeed, this can be quite exasperating for the CB because you could just have easily sent in some copies of *completed* forms as harder evidence, also some evidence of the retraining, such as new instructions to staff and training records.

Corrective Actions: Follow Up

If you have submitted corrective actions within the prescribed deadline but the CB is unhappy with either your corrective actions or the evidence to support them, they may offer you a final chance to submit further material. This is entirely at the CB's discretion. In some cases, it may be merely a clarification of material that you have sent in.

In Table 19.3, we have six nonconformities raised during an audit.

Table 19.3   Example of six minor nonconformities.

No.	Requirement ref.	Detail of nonconformity
1	2.10.1	Sieves were not being checked for each shift as per procedure – records show daily checks.
2	2.14.1	No records to show that HACCP was reviewed in May 2005.
3	3.3.1	No procedure in place to prevent changes to computerised records by unauthorised personnel.
4	3.9.2	No regular testing of traceability from raw material to finished product.
5	4.9.2.1	Unauthorised knife in process area.
6	7.2.5	No documented procedure to control use of personal medicines to prevent possible product contamination.

Example

For each of the nonconformities in Table 19.3, the manufacturer had completed the corrective action sheet and supplied evidence. However, for three of the nonconformities, the evidence supplied was insufficient. Their responses for these three were as follows:

Clause 2.10.1
This will be highlighted to the mixers.
   There was no evidence supplied to demonstrate that action had been taken.

Clause 3.3.1
We need to look at the system and investigate whether it is possible to adjust the current system. *Deadline*: End of this month.
   There was no evidence supplied to support this.

Clause 3.9.2
In January, we had done traceability tests but without evidence. From now on, we will print out all records so that we can show we do regular traceability tests.
   There was no evidence supplied to support this.

None of these responses were sufficiently definitive for the CB to be able to close them out, especially as no evidence was supplied. The CB, therefore, decided to request evidence to support the corrective actions.

In the case of Clause 2.10.1, the CB asked for evidence of the instruction to the staff.

For Clause 3.3.1, the CB explained that again they needed to see a procedure as evidence.

For Clause 3.9.2, the CB had to request evidence of an actual traceability test (both forwards and backwards).

The manufacturer promptly supplied the required evidence and the CB was able to issue a certificate.

While a certificate was issued in this case, for the CB to have to do this follow-up exercise takes time and resources and could be avoided. In extreme cases, this process can take so long that all deadlines expire and no certificate is issued. Understandably, the manufacturer wants to get the corrective actions in as soon as possible, but before you submit them it is worth taking some time to check them through to ensure that you are giving the CB all they need to close out your nonconformities.

One of the previous examples concerned traceability. If you had not carried out a traceability test at all (likely to be a major nonconformity), the submission of a statement that you *intend* to carry out a test, even with a date quoted will not be sufficient. There should be plenty of time for you to carry out a traceability test and submit the results of it to the CB. In such a case, the results would include all the paperwork you have examined so that the CB can see the codes you have traced.

Regarding the submission of evidence it does not matter whether the nonconformity was major or minor, if the opportunity is there to completely close out it must be taken.

The evidence discussed so far has been textual. Photographs are very useful indeed, especially to illustrate any number of situations of a physical nature, from improved proofing and other repairs to staff clothing, new equipment and so on. Recently, short video clips have also started appearing, which is an excellent way to show something in action: a metal detector detecting and rejecting, for example.

Ways of Sending Evidence

Much of the corrective action that you submit is likely to be in the form of documents. However, with the ease of email and electronic data, there are now good opportunities to be more innovative with evidence that can assist the CB in making a judgement by supporting your corrective actions with other types of material. Electronic transfer of data can also help to speed up the process, especially if the alternative is by post from another country.

Scanned forms, showing them to be completed, perhaps handwritten by staff can be emailed just as easily as they can be faxed or photocopied and posted. Digital photographs or video clips lend themselves to being emailed.

Electronic documents can also be linked, thus corrective action sheets that describe the action taken can be linked to the evidence documents, which can be helpful for the CB.

Other Ways of Helping the CB

When you are sending in documentary evidence to support corrective actions, make sure that it is well presented and the clause number is indicated on it so that the CB can tie it together with the relevant nonconformity. If the document is a long one, such as an amended procedure, it is helpful to highlight the relevant, amended parts.

Example

A tomato cannery had three minor nonconformities as shown in Table 19.4.

Their corrective action was supported by documentary evidence that ran to many pages (their HACCP system) plus one photograph. They had highlighted those areas of text, which were specifically relevant to the issues so that the CB could home in on them and readily see whether they now met requirements. The photograph was of a new compressed air cleaning system for the cans.

The manufacturer had helpfully provided good evidence to help the CB close out the corrective actions.

Table 19.4 Example of three minor nonconformities.

No.	Requirement ref.	Detail of nonconformity
Minor		
1	2.3.2	References to legislation were out of date.
2	2.9.1	Critical limits of detection for metal detectors were incorrect in the HACCP plan. Also, the critical limit for the magnets was not documented in the HACCP plan.
3	4.10.6.1	The cans for concentrate were not cleaned by compressed air before filling as specified in the HACCP plan.

Returning to Your Site

If you have received a Grade C or C+, the CB will have to revisit your site to verify the corrective action.

Part Three

If you have major or minor nonconformities, resulting in a Grade A or B (or A+ or B+), then, of course, you must submit evidence of corrective action to the CB. However, it may still be necessary for the CB to return to site to completely verify that you now comply with the requirements. For example, where a company has a major nonconformity, such that corrective action necessitates structural changes, it may be that the only way to properly verify this is a return to site.

Example

A company making high-risk products (ready meals, pies and baked desserts) did not have complete physical segregation of high- and low-risk areas. This was mainly due to the nature of their site, which was relatively old and had not been custom-built. In other words, the layout was not ideal.

The company came up with a solution to the issue by re-routing the traffic. However, it was not possible to fully understand or verify their action from two-dimensional plans and other documents submitted. Therefore, a return to site was made and the corrective action was indeed considered sufficient to close out the related nonconformities.

If the Auditor considers that a return to site to verify corrective actions is likely, they will make you aware of this during the closing meeting. In the case of a Grade C, because time is of the essence, they should agree a date for the return visit with you at that time.

Root Cause

By now, you will be aware that along with your corrective actions you must also supply corrective action plans for the root cause or each nonconformity. The CB will probably provide you with a form that includes this along with your corrective actions.

We discussed in Chapter 4 the concept of root cause analysis and how you will need some tools to carry this out. We have also looked at the fact that you will have had to carry out your action plan no later than the subsequent audit or fall foul of Clause 1.1.10. Note that root cause analysis is also required for companies who remain in the continual development phase of the Enrolment Programme, however, because of the nature of the Programme, closure is not necessarily required.

Your analysis should be wide-ranging and action plans may be quite diverse depending on the nature of the nonconformity. So, let us look at some examples.

One nonconformity from Table 19.1 was as follows:

There was mould growth on the ceiling in the fish preparation area. (Clause 4.4.5)

The corrective action here was very comprehensive, including photographs of the clean ceiling and an improvement in the internal audit system. But have we got to the root cause of the issue? Remember that the root cause action plan should prevent the issue from recurring. In theory, the corrective action here goes a long way to doing that. Applying one of the possible techniques to this (the 5 whys) might go like this:

(1) Why was there mould growth on the ceiling? *Answer*: Poor air extraction in the area.
(2) Why is our extraction poor? *Answer*: Extractors not the right capacity.
(3) Why are our extractors insufficient? *Answer*: Not upgraded as workload increased over time.
(4) Why did we not upgrade our extractors? *Answer*: Financial resources were focussed on expanding the production equipment.

In this case, I think we have reached the answer in four steps: the extraction equipment needs to be upgraded and more resources will be required to do this. You do not have to use all your questions if you can agree that the answer has been found at any stage.

Root cause may often come down to financial resources being required. Alternatively (or additionally), it may also be to do with improving staff training, for example. If we look at a simple nonconformity such as this one from our examples in this chapter:

The pest control biologist did not sign on to site or complete a health questionnaire on xx/xx/xx (Clause 7.3.2).

This might be a one-off incident. The corrective action has already ensured that a letter has gone to the contractor to remind them to follow your procedures, so what more can you do? In this case, your root cause analysis might identify a training issue with your reception or security staff. Another potential root cause for issues will be lack of human resources such as where an internal audit programme has failed. Remember this example from Chapter 13.

Example

A cutter and packer of meats and poultry, which was part of a group, had lost several key members of staff since their previous audit. The group had placed a new Operations Manager in charge, who also had responsibility for quality and technical management. However, he had had no proper training in the requirements of the FSQMS, being essentially a butcher by training. When they had their audit some months after his appointment, many of their Quality Systems were dormant, including internal audit, which had not been done for over a year.

Part Three

While the nonconformity might be corrected by getting the audits back underway using existing staff, there would clearly be a human resource issue here to address in longer term, not to mention a training requirement.

Enrolment Programme

We have looked at how the Enrolment Programme works in Chapter 5. To remind you, there is no obligation to correct nonconformities if you wish to remain in the continual development phase. However, you do have to send in corrective action plans and the CB does have to review them for the report. Additionally, you do have to send in a root cause plan for review. However, there is no time limit on closing them out. This means that for those in the continual development phase, at your next audit you may not have to meet the requirements of Clause 1.1.10 however it is good to aim to do so.

Summary

As far as you are concerned, we have come to the end of the process for achieving a certificate. You have submitted all your corrective actions and root cause plans and had them accepted by the CB. Remember to make your corrective actions definitive and provide good evidence. If you are going for certification this time, all that remains is for the CB to review the audit and, hopefully, issue your certificate. So, in the next chapter, we discuss what happens next.

Quiz No. 14

What would happen if the CB asked you to follow up with more information to verify corrective action but this was not resolved in the required timescale because your Technical Manager was on holiday?

20 Certification and What Happens Next

In this chapter, we look at what happens after a company has achieved certification. For those in the continual development phase, you will measure your success by your progress from one audit to the next.

Key points in this chapter
Re-audit dates
Delayed audits
Seasonal production
Surveillance
Scope extension

Success!

Well done! You have closed out any nonconformities to the satisfaction of your CB. They have finalised and uploaded your report to the BRC Directory. Your success will finally be marked with a nice certificate and you may rightly be proud of a significant achievement for your company. Enjoy the moment.

You are now part of a scheme that is ongoing and you should not rest too much on your laurels. The certificate will have some details on it regarding dates and other required features as shown in Figure 20.1. It is important to note your re-audit due and certificate expiry dates. You should have a plan in mind for your re-audit, for example ensure that key personnel will not take holidays during that time. Follow the Protocol and plan to have a re-audit in good time before the re-audit date. Remember that this can be done up to 28 days before the re-audit due date.

In this chapter, we look at life after certification.

Let us now look at some final parts of the Protocol that are concerned with continued certification.

Part Three

The BRC Global Standard for Food Safety: A Guide to a Successful Audit, Second Edition. Ron Kill
© 2012 John Wiley & Sons, Ltd. Published 2012 by John Wiley & Sons, Ltd.

CB Certification

This is to certify that

Sea Foods Ltd
BRC site code: 1234567
Fish St, Grimsby
GY1 1AB

Has been evaluated by CB Certification Ltd (a national accreditation body) accredited Certification Body No. 1234 and found to meet the requirements of

The Global Standard for Food Safety
Issue 6: July 2011

Scope:	Fresh, chilled and frozen fish portions and breaded fish products in flow-pack retail packs and 5-kg cases for catering customers.
Exclusions:	None
Product categories:	4: Raw fish and fish preparations
Achieved grade:	**B**
Audit programme:	Announced
Date of audit:	2nd August 2011
Reference:	XY036/011
Auditor number:	567007
Certificate issue date:	10th September 2011
Re-audit due date:	From 5th July 2012 to 2nd August 2012
Certificate expiry date:	13th September 2012

Authorised by:

C Brent

Cyril Brent, Director CB Certification Ltd.

CB Certification Ltd, 231 Basin Street, York YK5 6AB
This certificate remains the property of CB Certification Ltd

Figure 20.1 Example certificate.

Paragraph 13: BRC Logos and Plaques

You may now have a BRC logo to appear on your letter heading or other literature. You must apply to the BRC for this. Note that you may not use this logo or any reference to your Global Standard certification on products. You may also have a wall plaque from the BRC.

A UKAS-accredited CB may also issue you with a logo for use on your letter heading. The CB should inform you if they have such a logo. For other countries, you should check with your CB on this. If issued, this logo must be used correctly within certain rules and the CB should inform you on those. The CB will check on the use of this logo during subsequent audits.

Paragraph 14: The BRC Global Standards Directory

This subject is dealt with in Chapter 8.

Paragraph 15: Surveillance of Certificated Companies

It has always been the case that a CB could make a visit to a certificated company at any time as a surveillance visit. This is newly stated here for Issue 6. A new idea is that the BRC themselves may now carry out an audit if it is considered appropriate. Any nonconformities will be treated in the usual way and must be closed out within 28 days.

Note also that referrals, which may arise when a CB's performance is called into question (as mentioned in Chapter 3), may lead to such surveillance visits.

Paragraph 16: Ongoing Audit Frequency and Certification

Enjoy your success, but remember that your next audit will be in 12 months time (or maybe 6 months).

Paragraph 16.1: Scheduling re-audit dates

This part is about all the activities after your initial audit and how you maintain certification. In short, it is about the annual (or 6-monthly) re-audit process. Re-audits should always happen on or just before the anniversary of your original audit (unless you have a Grade C where it would be 6 months). There are some worked examples in Appendix 6 of the Standard to illustrate this. They show schedules for announced and unannounced audits.

It is important to note that it is *your responsibility*, not the CB's, to ensure that re-audits happen in the correct timescale. Most CBs will have a system in place for reminding you when your next audit is due; however, you should not rely on this.

Part Three

Paragraph 16.2: Delayed audits – justifiable circumstances

If you delay your audit for no good reason, then you will receive a major noncon-formity under Clause 1.1.8 of the requirements. Of course, it might be necessary to delay an audit when circumstances make it impossible, and there are certain justifiable circumstances for this that are listed in Clause 16.2. As you will see, they are extreme in nature and you will not be permitted to delay your audit for less important reasons, such as holidays, without the consequence of a major nonconformity before you even start your audit. If there is any doubt the CB may refer to the BRC for a judgment.

Paragraph 16.3: Audits undertaken prior to due dates

There is another important point to note here about the re-audit due date. There is a 28-day 'window' set out leading up to the final re-audit due date in which all subsequent audits must take place. That is to say you should not be asked by your CB to have the audit before that window, nor should you ask for it to be brought forward under normal circumstances. However, that is not to say that you, as a customer, cannot request an early re-audit at any time and there may be all sorts of reasons why that could be the case. However, if you do this, all your audit dates will be reset from the date of the new audit.

Paragraph 16.4: Seasonal production sites

There are difficulties for companies supplying seasonal products, especially when the season is very short. Clause 16.4 addresses this. One problem for short seasons hitherto has been that sometimes it was not possible to meet the re-audit date because the season was delayed or intermittent. The canning of tomatoes is a good example of this where a rainy season means that production depends not only on the general harvest period but on the day-to-day weather conditions.

Because we must carry out the audit when production is happening, in cases like this we must now do it as soon as possible, but if this goes beyond the due date, then any new certificate awarded is dated from the actual date of audit. This is the only instance of this being permissible in the Protocol.

The shortness of the season may mean that production could be over by the time that corrective actions are received by the CB.

Note: For any critical or major nonconformities, this is not permitted; they must be closed and verified during the season.

Also, if a company supplying seasonal products achieves a Grade C certificate, it does mean that the certificate naturally lapses before the following season.

If your site produces all year round but has seasonal variation of the products, it may be necessary to be visited more than once by the CB.

Note: You may have an unannounced audit as a seasonal supplier (see Chapter 5).

Paragraph 17: Communication with CBs

Remember that as part of the scheme you must also continue to follow the require-ments of the Standard. For example, you must keep your CB informed of any incidents such as a product recall. This is important in that it is your responsibility to let your CB know of any significant incident such as legal proceedings or a recall. In the latter case, should the Auditor find that you have not done this, you are likely to receive a nonconformity under Clause 3.11.4 of the requirements (which stipulates that you will let them know within three working days).

Once the CB is aware of any issues, they will review the situation to consider your continued certification. If they feel it is necessary, they will make a surveillance visit.

Paragraph 18: Appeals

This concerns appeals. If you are not happy with the final outcome of the certification decision, you should note that any appeals to your CB against a certification decision must be in writing and that you only have 7 days in which to do this. For their part, the CB has to have a documented procedure that details how they handle such appeals that you as a company are entitled to see. The CB in turn has 30 days to finalise an appeal.

Other Matters

This brings us to the end of the Protocol as such. Let us just look at some other points that might arise after you have your certificate.

Extensions to the scope

One of the things that might happen is that you wish to extend the scope of your certification before your re-audit is due. We briefly looked at the formality of extending scope in Chapter 6 when discussing Appendix 8 of the Standard. If you wish to extend your scope after you have been audited, you must notify your CB. They will then assess whether they need to audit your site again. If the extension is in keeping with your existing product range, the CB may take the view that they can extend your scope without a re-audit.

> **Example**
>
> A bakery site produces naan breads and they are the only named products in their scope. They wish to produce other flat breads on the same equipment. There are no additional ingredients involved; therefore the CB takes the view that no new risks are introduced and thus a new audit is not required.

However, if the extension to scope means that potential new hazards have been introduced, the CB is likely to require a revisit to site. The nature of this extension audit will be the same as a normal audit except that it may not be necessary to revisit every aspect of your existing scope. Thus, the duration could be shorter. Other differences are:

(1) Any nonconformities raised will not affect the existing grade (except see point 3).
(2) A report will be issued but not in the full standard format.
(3) If significant issues are found such as a critical nonconformity, a full re-audit of all systems and production will be arranged.
(4) A new certificate is issued with the new scope but the dates of re-audit remain the same as before.

So, all being well, you will receive a new certificate, but this will be dated using the same expiry date as before. The grade will also be unchanged even if the CB finds more nonconformities during this audit; however, they must be corrected in the normal way within the 28-day timescale.

Summary

We come to the end of this book. Achievement of a certificate for the Global Standard for Food Safety has shown that you and your company are committed to quality and continuous improvement. You are now committed to maintaining and improving your quality system. There is no looking back: the times that we live in dictate that we must all strive to improve and you must reconcile yourself to this and always be preparing for your next audit.

Quiz No. 15

Should the CB consider a new audit in the following cases?

(1) A bakery site produces standard pizza bases and wishes to add a gluten-free range to their scope.
(2) A site produces a range of soft, carbonated drinks packed in PET bottles. They have built a new extension where they are now bottling alcoholic drinks on contract. They wish to add this to their scope.
(3) A sandwich company who are certificated only for sandwiches wish to add filled wraps to their scope.

Part Three

References

BBC News (2010), Doubts raised over child food allergy rise. 10 August 2010. Available at www.bbc.co.uk/news/health-10925371.

BRC Global Standard for Storage and Distribution (2010), Issue 2, The Stationery Office (TSO), Norwich, www.tsoshop.co.uk.

BRC Best Practice Guideline – Product Recall Issue 2 (2007), The Stationery Office (TSO), Norwich, www.tsoshop.co.uk.

BRC Best Practice Guideline – Product Recall Issue 2 (2008), Internal audit, The Stationery Office (TSO), Norwich, www.tsoshop.co.uk.

BRC Best Practice Guideline – Product Recall Issue 2 (2008), Complaint handling, The Stationery Office (TSO), Norwich, www.tsoshop.co.uk.

BRC Best Practice Guideline – Product Recall Issue 2 (2008), Traceability, The Stationery Office (TSO), Norwich, www.tsoshop.co.uk.

BRC Best Practice Guideline – Product Recall Issue 2 (2008), Pest control, The Stationery Office (TSO), Norwich, www.tsoshop.co.uk.

BRC Best Practice Guideline – Product Recall Issue 2 (2008), Foreign body detection, The Stationery Office (TSO), Norwich, www.tsoshop.co.uk.

BRC Self Assessment Tool (2011), Available at http://www.brcglobalstandards.com/GlobalStandards/Guidelines/GuidanceDocumentation.aspx.

CNN (2010), Why are food allergies on the rise? 3 August 2010. Available at http://edition.cnn.com/2010/HEALTH/08/03/food.allergies.er.gut/index.html.

Codex Alimentarius Basic Food Texts on Food Hygiene (2003), Third Edition, Codex Alimentarius Commission, Joint FAO/WHO Food Standards Programme, FAO, Rome.

Directive 2003/89/EC amending directive 2000/13/EC of the European Parliament and Council of 20 March 2000 on the approximation of the laws of member states relating to the labelling, presentation and advertising of foodstuffs.

Journal of Food Protection (1998), **61** (9), 1246–1259, US National Advisory Committee on Microbiological Criteria for Foods: guidelines for application of HACCP principles.

Regulation (EC) No. 1924/2006 of the European Parliament and of the Council of 20 December 2006 on nutrition and health claims made on foods.

The Animal By-Products Regulations (2005), ISBN 0110732804, The Stationery Office (TSO), Norwich, www.tsoshop.co.uk.

The Weights and Measures (Packaged Goods) Regulations (2006), ISBN 011074330X, The Stationery Office (TSO), Norwich, www.tsoshop.co.uk.

Useful Websites

BRC Standards webpage: www.brcglobalstandards.com.
BRC Directory: www.brcdirectory.com.
Codex Alimentarius: www.codexalimentarius.net.
Food Standards Agency website: www.food.gov.uk.
GFSI: www.mygfsi.com.
TSO: www.tso.co.uk.
UKAS-accredited CBs: http://www.ukas.com/about_accreditation/accredited_bodies/
    certification_body_schedules.asp.
United Kingdom Accreditation Service (UKAS): www.ukas.com.

Appendices

Appendix 1 Answers to Quizzes and Exercise

Quiz No. 1 (Chapter 2)

Just one: 3.4 internal audit.

Quiz No. 2 (Chapter 4)

(1) 14
(2) It is likely to be audited during the document review.
(3) When carrying out any corrective action including from issues arising internally, customer complaints and from your BRC audit.
(4) 7

Quiz No. 3 (Chapter 5)

(1) Clauses used in Table 5.4:
    a. 3.9.2
    b. 4.4.5
    c. 4.4.9
    d. 4.8.3
    e. 4.9.3.4.2
    f. 6.1.2
    g. 7.2.1
(2) 10
(3) Clauses used in the 'Try This' exercise:
    a. 5.2.4
    b. 4.3.6
    c. 6.2.1
    d. 1.1.8
    e. 7.2.1

The BRC Global Standard for Food Safety: A Guide to a Successful Audit, Second Edition. Ron Kill
© 2012 John Wiley & Sons, Ltd. Published 2012 by John Wiley & Sons, Ltd.

f. 4.14.1
g. 4.13.8
h. 3.9.2

Quiz No. 4 (Chapter 6)

(1) Low risk.
(2) High care.
(3) 18
(4) No, it is outside the 30-mile limit.
(5) Withdraw the existing certificate and arrange for a new full audit.

Quiz No. 5 (Chapter 7)

(1) 2 days.
(2) More than 100.
(3) 49
(4) Yes.
(5) It is too vague in describing the products and should be much more specific. Also, words like 'sale' should not be used.

Quiz No. 6 (Chapter 11)

(1) A major nonconformity against Clause 1.1.8.
(2) Monthly.
(3) The person with overall responsibility for the site.
(4) Monitored and reported on at least quarterly (every 3 months).
(5) 1.1.10

Quiz No. 7 (Chapter 12)

(1) 4 (2.2, 2.12, 2.13, 2.14)
(2) Annually.
(3) 2.3.1
(4) No, 2.1.1 requires that they have in-depth knowledge.
(5) If they are subjective, they must be supported by guidance.

Quiz No. 8 (Chapter 13)

(1) 3.7.1 and 3.10.1.
(2) 12 months.
(3) Monthly.
(4) Every 3 years.
(5) 3.4, 3.7 and 3.9.

Quiz No. 9 (Chapter 14)

   (1) Enclosed product areas or open product areas when treating an active infestation.
   (2) Annually.
   (3) At least annually.
   (4) Magnetic field.
   (5) Time, detergent concentration, flow rate and temperatures.
   (6) Typically, every 4 hours.
   (7) Where the product is susceptible to weather damage.
   (8) Product that is fine filtered immediately before packing.
   (9) At access to production areas and other appropriate points.
  (10) False, only where there are high-care and high-risk areas.
  (11) For initial product cleaning.
  (12) Entry to high-risk zones.

Quiz No. 10 (Chapter 15)

   (1) At least 6-monthly.
   (2) False, you must prevent cross-contamination wherever possible.
   (3) When using analyses that are critical to product safety or legality.
   (4) Inform the supplier of any particular characteristics of the food or use to which it will be put by the consumer.
   (5) Two: 5.2.9 and 5.5.2.4.

Quiz No. 11 (Chapter 16)

   (1) That it has been suitably cleaned and that all products and packaging from the previous run are removed.
   (2) Reference measuring equipment.
   (3) A failure alert system.

Quiz No. 12 (Chapter 17)

   (1) Not any more.
   (2) A sample of each batch.
   (3) False, they may have pockets below the waist.
   (4) You, the company.
   (5) By audit or possession of relevant certification.

Quiz No. 13 (Chapter 18)

   (1) 28 days.
   (2) You, the client.
   (3) The CB.
   (4) After the technical review by the CB.

Quiz No. 14 (Chapter 19)

You still have to meet the deadline. The CB might give you a little leeway, but this is unlikely to extend beyond 42 days. There is a good chance you will not be certificated.

Quiz No. 15 (Chapter 20)

(1) Yes. This would require a whole new level of segregation on the part of the factory and the CB would need to see it.
(2) Yes. It is a different product with different legal requirements.
(3) Possibly not. The wraps would be a very similar process to sandwich-making.

'Try This' Exercise (Chapter 5)

(1) Minor. Although segregation for allergens is crucially important, this was just one scoop of the wrong colour that might not yet have led to a contamination issue itself, but potentially it could.
(2) Major. This is a clear failure to meet the clause. However, there is no evidence that unsafe product had actually resulted; hence, it is not a critical nonconformity.
(3) Critical. The company is failing to meet a legal requirement and has almost certainly supplied illegal products.
(4) Major. A clear failure to meet the clause.
(5) Minor. Only one member of staff was involved.
(6) Minor. A small quantity of product is on the floor, the rest by implication complied with requirements.
(7) Major. They have completely disregarded this requirement.
(8) Minor. They had carried out a test, so have partially met the clause, but they are a little late.

Appendix 2   Where Did Issue 5 Go?

The following table shows where all the clauses from Issue 5 went into Issue 6. Note that many of the Issue 6 versions are different and have been extensively reworded. However, that is where you will find the relevant subject matter if you are trying to track them.

Issue 5 clause	Issue 6 destination
Section 1	
1	1.1
1.1	1.1.5
1.2	Removed
1.3	1.1.2
1.4	1.1.4
1.5	1.1.3
1.6	1.1.3
1.7	1.1.3
1.8	1.1.3
1.9	1.1.3
1.10	1.1.7
1.11	1.1.8
1.12	1.1.9
1.13	1.1.10
Section 2	
2	2
2.1.1	2.1.1
2.1.2	2.1.1
2.1.3	2.1.1
2.1.4	Removed
2.2.1	2.3.1
2.2.2	2.3.2

(continued)

The BRC Global Standard for Food Safety: A Guide to a Successful Audit, Second Edition. Ron Kill
© 2012 John Wiley & Sons, Ltd. Published 2012 by John Wiley & Sons, Ltd.

Issue 5 clause	Issue 6 destination
Section 2	
2.2.3	2.3.1
2.3.1	2.4.1
2.4.1	2.5.1
2.5.1	2.6.1
2.6.1	2.7.1
2.6.2	2.7.2
2.6.3	2.7.3
2.7.1	2.8.1
2.8.1	2.9.1
2.8.2	2.9.1
2.8.3	2.9.2
2.9.1	2.10.1
2.9.2	2.10.1
2.9.3	2.10.2
2.10.1	2.11.1
2.10.2	Removed
2.11.1	2.12.1
2.11.2	2.12.1
2.12.1	2.13.1
2.13.1	2.14.1
2.13.2	2.14.1
Section 3	
3.1	1.1.1
3.1.1	1.1.1
3.2	3.1
3.2.1	3.1.1
3.2.2	3.1.2
3.3	1.2
3.3.1	1.2.1
3.3.2	1.2.1
3.3.3	1.2.1
3.3.4	1.2.2
3.3.5	1.1.6
3.4	Removed
3.4.1	Removed
3.4.2	Removed
3.4.3	Removed
3.4.4	Removed
3.5	3.4
3.4.1	3.5.1
3.4.2	3.5.2
3.5.3	3.4.4
3.5.4	3.4.4
3.5.5	3.4.4

Issue 5 clause	Issue 6 destination
Section 3	
3.5.6	3.4.4
3.6	3.5.1 and 3.5.3
3.6.1	3.5.1.2 and 3.5.3.1
3.6.2	3.5.1.2
3.6.3	3.5.1.3
3.6.4	Removed
3.7.1	3.2
3.7.1.1	3.2
3.7.1.2	3.1.3
3.7.1.3	3.2.1
3.7.1.4	3.2.1
3.7.2	3.6
3.7.2.1	3.6.1
3.7.2.2	3.6.2
3.7.2.3	3.6.4
3.7.2.5	3.6.5
3.7.3	3.3
3.7.3.1	3.3.1
3.7.3.2	3.3.1
3.7.3.3	Removed
3.7.3.4	3.3.2
3.7.3.5	3.3.2
3.8	3.7
3.8.1	3.7.1
3.8.2	3.7.1
3.8.3	3.7.1
3.8.4	3.7.1
3.9	3.9
3.9.1	3.9.1
3.9.2	3.9.2
3.9.3	5.3.2
3.9.4	3.9.3
3.10	3.10
3.10.1	3.10.1
3.10.2	3.10.1
3.10.3	3.10.2
3.11	3.11
3.11.1	3.11.1
3.11.2	3.11.2
3.11.3	3.11.2
3.11.4	3.11.2
3.11.5	3.11.3
3.11.6	3.11.3
3.11.7	3.11.4

(continued)

Issue 5 clause	Issue 6 destination
Section 4	
4.1	4.1
4.1.1	4.1.1
4.1.2	4.1.2
4.1.3	Removed
4.1.4	4.1.2
4.1.5	4.1.3
4.2	4.2
4.2.1	4.2.2
4.2.2	4.2.2
4.2.3	4.2.2
4.2.4	4.2.1
4.2.5	Removed
4.2.6	4.2.3
4.3.1	4.3
4.3.1.1	4.3.4
4.3.1.2	4.3.4
4.3.1.3	4.3.4
4.3.1.4	Removed
4.3.1.5	4.3.7
4.3.1.6	4.3.1.5
4.3.1.7	4.3.8
4.3.1.8	4.3.5
4.3.1.9	4.3.6
4.3.1.10	4.3.5
4.3.2	4.4
4.3.2.1	4.4.1
4.3.2.2.1	4.4.2
4.3.2.2.2	4.4.3
4.3.2.2.3	4.4.3
4.3.2.3.1	4.4.5
4.3.2.3.2	4.4.6
4.3.2.4.1	4.4.7
4.3.2.4.2	4.4.8
4.3.2.5.1	4.4.9
4.3.2.5.2	4.4.9
4.3.2.6.1	4.4.10
4.3.2.6.2	4.4.11
4.3.2.7.1	4.4.12
4.3.2.7.2	4.4.13
4.3.2.7.3	4.4.13
4.4	4.5
4.4.1	4.5.1
4.4.2	4.5.4
4.5	4.6
4.5.1	4.6.1
4.5.2	Removed

Issue 5 clause	Issue 6 destination
Section 4	
4.5.3	4.6.2
4.6	4.7.1
4.6.1	4.7
4.6.2	4.7.1
4.6.3	4.7.4
4.6.4	4.7.2
4.6.5	4.7.3
4.6.6	4.3.3
4.6.7	4.7.4
4.6.8	4.7.5
4.6.9	4.7.6
4.7	4.8
4.7.1	4.3.2/4.8.1
4.7.2	4.8.2
4.7.3	4.8.3
4.7.4	4.8.6
4.7.5	4.8.7
4.7.6	4.8.8
4.7.7	4.8.9
4.7.8	4.8.10
4.7.9	4.8.9
4.7.10	Removed
4.7.11	4.8.5
4.8	4.9
4.8.1	Removed
4.8.2	4.9.1.1
4.8.3.1	4.9.2.1
4.8.3.2	4.9.2.1
4.8.3.3	Removed
4.8.3.4	4.9.2.2
4.8.4.1	4.9.3.1
4.8.4.2	4.9.3.2
4.8.4.3	4.9.3.3
4.8.5.1	4.9.4.1
4.8.6.1	4.10.2.2
4.9	4.11
4.9.1	4.11.1
4.9.2	4.11.6.1
4.9.3	4.11.3
4.9.4	4.11.5
4.9.5	4.11.2
4.9.6	Removed
4.10	4.12
4.10.1	4.12
4.10.2	4.12.1

(continued)

Issue 5 clause	Issue 6 destination
Section 4	
4.10.3	4.12.1
4.10.4	4.12.3
4.10.5	4.12.4
4.11	4.13
4.11.1	Removed
4.11.2	4.13.1
4.11.3	4.13.3
4.11.4	4.13.4
4.11.5	4.13.5
4.11.6	4.13.6
4.11.7	4.13.7
4.11.8	4.13.9
4.12	4.14 and 4.15
4.12.1	4.15.1
4.12.2	4.14.2
4.12.3	4.15.4
4.12.4	4.14.4
4.12.5	4.14.5
4.12.6	4.15.7
4.12.7	4.15.2
4.12.8	4.15.5
4.12.9	4.15.6
Section 5	
5.1	5.1
5.1.1	5.1.2
5.1.2	5.1.3
5.1.3	5.1.4
5.1.4	Removed
5.1.5	5.4.1
5.1.6	Removed
5.1.7	5.1.6
5.1.8	Removed
5.2	5.2
5.2.1.1	5.2.1
5.2.1.2	5.2.2
5.2.1.3	5.2.3
5.2.1.4	5.2.5
5.2.1.5	5.2.7
5.2.1.6	5.2.8
5.2.1.7	5.2.9
5.2.1.8	5.2.10
5.2.2.1	5.3.1
5.2.2.2	5.3.3
5.3.1	4.10.3.1 and 4.10.3.2
5.3.2	4.10.1.2

Issue 5 clause	Issue 6 destination
Section 5	
5.3.3	4.10.3.3
5.3.4	4.10.3.4
5.3.5	4.10.3.6
5.4	5.4
5.4.1	5.4.1
5.4.2	5.4.2
5.4.3	5.4.2
5.4.4	5.4.3
5.4.5	Removed
5.5	5.5
5.5.1.1	5.5.1.1
5.5.1.2	5.5.1.2
5.5.1.3	5.5.1.3
5.5.1.4	5.5.1.4
5.5.2.1	5.5.2.1
5.5.2.2	5.5.2.2
5.5.2.3	5.5.2.3
5.5.2.4	5.5.2.4
5.6	3.8
5.6.1	3.8.1
5.6.2	3.8.1
5.6.3	3.8.1
5.7	5.6
5.7.1	5.6.1
Section 6	
6	Removed
6.1	6.1
6.1.1	6.1.1
6.1.2	6.1.2
6.1.3	Removed
6.1.4	6.1.3
6.1.5	6.1.5
6.1.6	Removed
6.1.7	6.1.7
6.1.8	Removed
6.2	6.2
6.2.1	6.2.1
6.2.2	6.2.2
6.3	6.3
6.3.1	6.3.1
6.3.2	6.3.2
6.3.3	6.3.1
6.3.4	6.3.4

(continued)

Issue 5 clause	Issue 6 destination
Section 7	
7.1	7.1
7.1.1	7.1.1
7.1.2	7.1.2
7.1.3	7.1.3
7.1.4	7.1.4
7.1.5	7.1.5
7.2	4.3.2
7.2.1	4.3.2
7.2.2	4.3.2
7.2.3	4.3.2
7.2.4	4.3.3
7.3	7.2
7.3.1	7.2.1
7.3.2	7.2.1
7.3.3	7.2.1
7.3.4	7.2.2
7.3.5	7.2.1
7.3.6	7.2.1
7.3.7	4.8.8
7.3.8	7.2.3
7.3.9	7.2.4
7.3.10	7.2.5
7.4	7.3
7.4.1	7.3.1
7.4.2	7.3.2
7.4.3	7.3.3
7.5	7.4
7.5.1	7.4.1
7.5.2	7.4.2
7.5.3	4.8.3
7.5.4	7.4.3
7.5.5	Removed
7.5.6	7.4.2
7.5.7	7.4.2
7.5.8	Removed
7.5.9	7.4.5
7.5.10	4.8.4

Appendix 3 New Clauses for Issue 6

The following table shows the completely new clauses for Issue 6.

Clause 1	Clause 2	Clause 3	Clause 4	Clause 5	Clause 6	Clause 7
None	2.2	3.4.4	4.3.1	5.1.1	6.1.4	7.4.4
		3.5.1.1	4.4.4	5.1.5	6.1.6	7.4.6
		3.5.2	4.5.2	5.2.4	6.3.3	
		3.5.2.1	4.5.3	5.2.6		
		3.5.3	4.8.4			
		3.5.3.1	4.9.1.2			
		3.5.3.2	4.9.3.4.1			
		3.5.4	4.9.3.4.2			
		3.5.4.1	4.9.3.4.3			
		3.5.4.2	4.10			
		3.5.4.3	4.10.1.1			
		3.5.4.4	4.10.1.2			
			4.10.1.3			
			4.10.1.4			
			4.10.2.1			
			4.10.3.5			
			4.10.4.1			
			4.10.5.1			

(continued)

The BRC Global Standard for Food Safety: A Guide to a Successful Audit, Second Edition. Ron Kill
© 2012 John Wiley & Sons, Ltd. Published 2012 by John Wiley & Sons, Ltd.

Clause 1	Clause 2	Clause 3	Clause 4	Clause 5	Clause 6	Clause 7
			4.10.6.2			
			4.11.6.2			
			4.11.6.3			
			4.12.2			
			4.13.2			
			4.13.8			
			4.14.3			
			4.15			

Note: There are also several clauses that have been greatly revised or augmented for Issue 6, which are described in the relevant chapter. An example is 3.6.3 concerning finished product specifications. They were referred to in the previous SOI and were implicit in the clauses; however, the extra detail and new focus on branded product specifications make this almost a new clause. Another example is 4.4.13 with the specific emphasis on positive air pressure and number of air changes for high-risk zones. There are others.

Appendix 4   Changes to Product Categories from Issue 5 to Issue 6

Issue 5 reference	Issue 6 reference	Category	Changes
Appendix 3 – product categories	Appendix 4 – product categories	3	High-care/low-risk principles omitted. Comminuted fish products added from Category 4.
		4	High-care/low-risk principles omitted. Comminuted fish products moved to Category 3, for example fish pie omitted.
		5	Storage conditions now only 'fresh'.
		7	Butter added, cheese food added.
		8	Hot and cold eating pies removed to Category 10.
		10	Hot and cold eating pies added.
		12	Milk and cereal beverages omitted.
		14	'Frozen' added to storage conditions.
		15	Sugar and flour added. Gases omitted.
		16	Sugar removed.
		17	Now includes 'Porridge' oats.
		18	Butter moved to Category 7.

The BRC Global Standard for Food Safety: A Guide to a Successful Audit, Second Edition. Ron Kill
© 2012 John Wiley & Sons, Ltd. Published 2012 by John Wiley & Sons, Ltd.

Index

The BRC Global Standard for Food Safety: A Guide to a Successful Audit, Second Edition. Ron Kill
© 2012 John Wiley & Sons, Ltd. Published 2012 by John Wiley & Sons, Ltd.

Food Science and Technology

GENERAL FOOD SCIENCE & TECHNOLOGY, ENGINEERING AND PROCESSING

Organic Production and Food Quality: A Down to Earth Analysis	Blair	9780813812175
Handbook of Vegetables and Vegetable Processing	Sinha	9780813815411
Nonthermal Processing Technologies for Food	Zhang	9780813816685
Thermal Procesing of Foods: Control and Automation	Sandeep	9780813810072
Innovative Food Processing Technologies	Knoerzer	9780813817545
Handbook of Lean Manufacturing in the Food Industry	Dudbridge	9781405183673
Intelligent Agrifood Networks and Chains	Bourlakis	9781405182997
Practical Food Rheology	Norton	9781405199780
Food Flavour Technology, 2nd edition	Taylor	9781405185431
Food Mixing: Principles and Applications	Cullen	9781405177542
Confectionery and Chocolate Engineering	Mohos	9781405194709
Industrial Chocolate Manufacture and Use, 4th edition	Beckett	9781405139496
Chocolate Science and Technology	Afoakwa	9781405199063
Essentials of Thermal Processing	Tucker	9781405190589
Calorimetry in Food Processing: Analysis and Design of Food Systems	Kaletunç	9780813814834
Fruit and Vegetable Phytochemicals	de la Rosa	9780813803203
Water Properties in Food, Health, Pharma and Biological Systems	Reid	9780813812731
Food Science and Technology (textbook)	Campbell-Platt	9780632064212
IFIS Dictionary of Food Science and Technology, 2nd edition	IFIS	9781405187404
Drying Technologies in Food Processing	Chen	9781405157636
Biotechnology in Flavor Production	Havkin-Frenkel	9781405156493
Frozen Food Science and Technology	Evans	9781405154789
Sustainability in the Food Industry	Baldwin	9780813808468
Kosher Food Production, 2nd edition	Blech	9780813820934

FUNCTIONAL FOODS, NUTRACEUTICALS & HEALTH

Functional Foods, Nutraceuticals and Degenerative Disease Prevention	Paliyath	9780813824536
Nondigestible Carbohydrates and Digestive Health	Paeschke	9780813817620
Bioactive Proteins and Peptides as Functional Foods and Nutraceuticals	Mine	9780813813110
Probiotics and Health Claims	Kneifel	9781405194914
Functional Food Product Development	Smith	9781405178761
Nutraceuticals, Glycemic Health and Type 2 Diabetes	Pasupuleti	9780813829333
Nutrigenomics and Proteomics in Health and Disease	Mine	9780813800332
Prebiotics and Probiotics Handbook, 2nd edition	Jardine	9781905224524
Whey Processing, Functionality and Health Benefits	Onwulata	9780813809038
Weight Control and Slimming Ingredients in Food Technology	Cho	9780813813233

INGREDIENTS

Hydrocolloids in Food Processing	Laaman	9780813820767
Natural Food Flavors and Colorants	Attokaran	9780813821108
Handbook of Vanilla Science and Technology	Havkin-Frenkel	9781405193252
Enzymes in Food Technology, 2nd edition	Whitehurst	9781405183666
Food Stabilisers, Thickeners and Gelling Agents	Imeson	9781405132671
Glucose Syrups – Technology and Applications	Hull	9781405175562
Dictionary of Flavors, 2nd edition	De Rovira	9780813821351
Vegetable Oils in Food Technology, 2nd edition	Gunstone	9781444332681
Oils and Fats in the Food Industry	Gunstone	9781405171212
Fish Oils	Rossell	9781905224630
Food Colours Handbook	Emerton	9781905224449
Sweeteners Handbook	Wilson	9781905224425
Sweeteners and Sugar Alternatives in Food Technology	Mitchell	9781405134347

FOOD SAFETY, QUALITY AND MICROBIOLOGY

Food Safety for the 21st Century	Wallace	9781405189118
The Microbiology of Safe Food, 2nd edition	Forsythe	9781405140058
Analysis of Endocrine Disrupting Compounds in Food	Nollet	9780813818160
Microbial Safety of Fresh Produce	Fan	9780813804163
Biotechnology of Lactic Acid Bacteria: Novel Applications	Mozzi	9780813815831
HACCP and ISO 22000 – Application to Foods of Animal Origin	Arvanitoyannis	9781405153669
Food Microbiology: An Introduction, 2nd edition	Montville	9781405189132
Management of Food Allergens	Coutts	9781405167581
Campylobacter	Bell	9781405156288
Bioactive Compounds in Foods	Gilbert	9781405158756
Color Atlas of Postharvest Quality of Fruits and Vegetables	Nunes	9780813817521
Microbiological Safety of Food in Health Care Settings	Lund	9781405122207
Food Biodeterioration and Preservation	Tucker	9781405154178
Phycotoxins	Botana	9780813827001
Advances in Food Diagnostics	Nollet	9780813822211
Advances in Thermal and Non-Thermal Food Preservation	Tewari	9780813829685

For further details and ordering information, please visit www.wiley.com/go/food

Food Science and Technology from Wiley-Blackwell

SENSORY SCIENCE, CONSUMER RESEARCH & NEW PRODUCT DEVELOPMENT

Title	Author	ISBN
Sensory Evaluation: A Practical Handbook	Kemp	9781405162104
Statistical Methods for Food Science	Bower	9781405167642
Concept Research in Food Product Design and Development	Moskowitz	9780813824246
Sensory and Consumer Research in Food Product Design and Development	Moskowitz	9780813816326
Sensory Discrimination Tests and Measurements	Bi	9780813811116
Accelerating New Food Product Design and Development	Beckley	9780813808093
Handbook of Organic and Fair Trade Food Marketing	Wright	9781405150583
Multivariate and Probabilistic Analyses of Sensory Science Problems	Meullenet	9780813801780

FOOD LAWS & REGULATIONS

Title	Author	ISBN
The BRC Global Standard for Food Safety: A Guide to a Successful Audit	Kill	9781405157964
Food Labeling Compliance Review, 4th edition	Summers	9780813821818
Guide to Food Laws and Regulations	Curtis	9780813819464
Regulation of Functional Foods and Nutraceuticals	Hasler	9780813811772

DAIRY FOODS

Title	Author	ISBN
Dairy Ingredients for Food Processing	Chandan	9780813817460
Processed Cheeses and Analogues	Tamime	9781405186421
Technology of Cheesemaking, 2nd edition	Law	9781405182980
Dairy Fats and Related Products	Tamime	9781405150903
Bioactive Components in Milk and Dairy Products	Park	9780813819822
Milk Processing and Quality Management	Tamime	9781405145305
Dairy Powders and Concentrated Products	Tamime	9781405157643
Cleaning-in-Place: Dairy, Food and Beverage Operations	Tamime	9781405155038
Advanced Dairy Science and Technology	Britz	9781405136181
Dairy Processing and Quality Assurance	Chandan	9780813827568
Structure of Dairy Products	Tamime	9781405129756
Brined Cheeses	Tamime	9781405124607
Fermented Milks	Tamime	9780632064588
Manufacturing Yogurt and Fermented Milks	Chandan	9780813823041
Handbook of Milk of Non-Bovine Mammals	Park	9780813820514
Probiotic Dairy Products	Tamime	9781405121248

SEAFOOD, MEAT AND POULTRY

Title	Author	ISBN
Handbook of Seafood Quality, Safety and Health Applications	Alasalvar	9781405180702
Fish Canning Handbook	Bratt	9781405180993
Fish Processing – Sustainability and New Opportunities	Hall	9781405190473
Fishery Products: Quality, safety and authenticity	Rehbein	9781405141628
Thermal Processing for Ready-to-Eat Meat Products	Knipe	9780813801483
Handbook of Meat Processing	Toldra	9780813821825
Handbook of Meat, Poultry and Seafood Quality	Nollet	9780813824468

BAKERY & CEREALS

Title	Author	ISBN
Whole Grains and Health	Marquart	9780813807775
Gluten-Free Food Science and Technology	Gallagher	9781405159159
Baked Products – Science, Technology and Practice	Cauvain	9781405127028
Bakery Products: Science and Technology	Hui	9780813801872
Bakery Food Manufacture and Quality, 2nd edition	Cauvain	9781405176132

BEVERAGES & FERMENTED FOODS/BEVERAGES

Title	Author	ISBN
Technology of Bottled Water, 3rd edition	Dege	9781405199322
Wine Flavour Chemistry, 2nd edition	Bakker	9781444330427
Wine Quality: Tasting and Selection	Grainger	9781405113663
Beverage Industry Microfiltration	Starbard	9780813812717
Handbook of Fermented Meat and Poultry	Toldra	9780813814773
Microbiology and Technology of Fermented Foods	Hutkins	9780813800189
Carbonated Soft Drinks	Steen	9781405134354
Brewing Yeast and Fermentation	Boulton	9781405152686
Food, Fermentation and Micro-organisms	Bamforth	9780632059874
Wine Production	Grainger	9781405113656
Chemistry and Technology of Soft Drinks and Fruit Juices, 2nd edition	Ashurst	9781405122863

PACKAGING

Title	Author	ISBN
Food and Beverage Packaging Technology, 2nd edition	Coles	9781405189101
Food Packaging Engineering	Morris	9780813814797
Modified Atmosphere Packaging for Fresh-Cut Fruits and Vegetables	Brody	9780813812748
Packaging Research in Food Product Design and Development	Moskowitz	9780813812229
Packaging for Nonthermal Processing of Food	Han	9780813819440
Packaging Closures and Sealing Systems	Theobald	9781841273372
Modified Atmospheric Processing and Packaging of Fish	Otwell	9780813807683
Paper and Paperboard Packaging Technology	Kirwan	9781405125031

For further details and ordering information, please visit www.wiley.com/go/food